PRAISE FOR *OCEAN OF INSIGHT*

"An authentic and exciting adventure of oceanic and psychological depth that each one of us should venture on at least once in their lifetime. With engaging details of an experienced sailor and of a seasoned meditator, Heather Mann invites us on a human journey of courage, challenge, and spiritual wonder. She artfully weaves her own meditative practice and reflections with the fabric of a classic quest to find meaning and purpose for herself and human kind in our relationship with Mother Earth. She shows us that with the practice of mindfulness, one can find coolness for oneself and thus can help find coolness for the Earth."

—Bhikshu Thich Chan Phap Dung, Dharma Teacher of the Plum Village Tradition

"Captivating and insightful, Heather Mann's *Ocean of Insight* captures her love of the natural world, her knowledge of environmental science, and her practice of mindfulness and inner knowing. She expertly narrates her riveting adventures with husband Dave aboard *Wild Hair*, their 44-foot sailboat, and she simultaneously draws us in the bountiful insights she gleans throughout their treacherous journeys across the wilderness of the Atlantic. It is a compelling story of ecological mindfulness and expanded awareness of an evolving human consciousness that has the capacity to bring forth healing and transformation and to find a new way of living and being on the earth. You can't help but be changed for the better for having read this book."

—Carolyn Rivers, Founder and Director, The Sophia Institute

OCEAN OF INSIGHT

OCEAN OF INSIGHT

A Sailor's Voyage
from Despair to Hope

HEATHER LYN MANN

**PARALLAX
PRESS**

BERKELEY, CALIFORNIA

Parallax Press
P.O. Box 7355
Berkeley, California 94707
parallax.org

Parallax Press is the publishing division of
Unified Buddhist Church, Inc.
© 2016 Heather Lyn Mann
All rights reserved
Printed in The United States of America

Cover and text design by Debbie Berne
Maps and illustrations © John Barnett
Author photo © Sara Shaffer

The names of fellow sailors and ships are changed to safeguard their privacy. All other people, places, and events
are real.

Library of Congress Cataloging-in-Publication Data is
available upon request.

ISBN: 978-1-941529-30-0

1 2 3 4 5 / 20 19 18 17 16

To all that is magnificent
Including the whale
And especially Dave

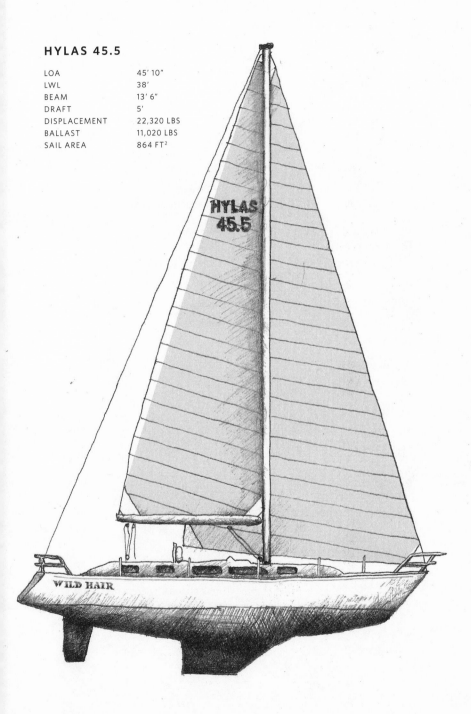

HYLAS 45.5

LOA	45' 10"
LWL	38'
BEAM	13' 6"
DRAFT	5'
DISPLACEMENT	22,320 LBS
BALLAST	11,020 LBS
SAIL AREA	864 FT²

CONTENTS

PART III
UNLEARNING WHAT IS KNOWN

13
Beginner's Mind 329

GLOSSARY OF SAILING TERMS

Aft: at or toward the back end of a boat

Batten: a long, narrow brace inserted into a pocket in the mainsail to retain optimal sail shape

Boom: the metal pole running horizontally at the base of a mainsail

Bow: the most forward, pointed part of a boat

Cleat: a metal fitting with two horns located on a deck or pier around which lines are attached

Cockpit: a protected space above decks from which a boat is steered

Dinghy: 1. a small, inflatable boat used to tender crew from ship to shore. 2. a gray and white American shorthair cat who lives aboard sailing vessel *Wild Hair.*

Galley: a ship's kitchen

Head: a ship's toilet

Heel: the degree of angle a sailboat tilts away from the wind

Helm: the location of the wheel and other equipment used to steer a ship

Hull: the outer shell of a boat, including the bottom, sides, and deck

Jib: the most forward, triangular sail on a sloop

Jibing: altering course while sailing downwind and making the wind shift from one side of the boat to the other

Keel: the heavy fin attached to the lowest point of the hull that counteracts heel and the sideways pressure of wind

Line: a ship's rope

Mainsail: the principal, triangular sail on the mast of a sloop

Mast: the vertical, metal stick that suspends the sails

Point of Sail: various names for a traveling sailboat relative to the angle of wind, including: close haul, close reach, beam reach, broad reach, and running

Port: at or toward the left side of the boat when facing forward

Rudder: the blade attached to the stern hull, controlled by the helm, that steers the ship

Sloop: a single-masted sailboat equipped with mainsail and jib

Starboard: at or toward the right side of the boat when facing forward

Stern: the back end of the boat

Tacking: altering course while sailing upwind and making the wind shift from one side of the boat to the other

Transom: a horizontal platform located at a ship's very stern

Trim: the act of manipulating lines to change the shape and position of the sails

Winch: a drum-shaped machine, turned with a handle or crank, used for hauling or hoisting line

PART I
TRANSMISSION OF LEARNING

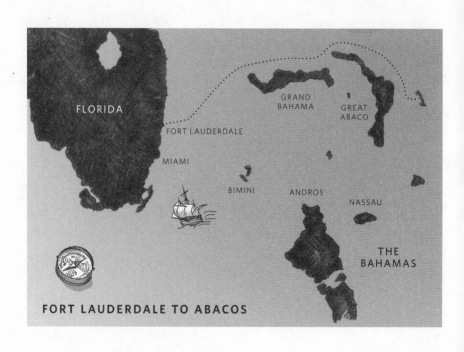

FLORIDA

FORT LAUDERDALE

MIAMI

BIMINI

GRAND
BAHAMA

GREAT
ABACO

ANDROS

NASSAU

THE
BAHAMAS

FORT LAUDERDALE TO ABACOS

1 FEAR

I look upon the jagged shore to calculate the time until impact. It's difficult to know exactly because the anchors scrape the ocean floor, slowing
our approach. The storm is building. Waves slam against the bow and
drive us backward. The ship's engine picked this moment to stop func-
tioning, so Dave and I are suddenly, inexplicably, without power. The
sun is slipping low and soon we will be without light.

I sailed my ship, *Wild Hair,* to this spot because I wanted lobster
from the reef for a New Year's dinner. But this is a place of peril in a
gale—especially with a busted throttle cable. Now I am exposed, disabled, at risk of losing my ship and maybe my life.

A primal panic starts simmering at the base of my spine. It wraps my
intestines. My limbs feel thick as logs and my thoughts are slow; they
roll into consciousness with the speed of old movie credits. Usually, I'm a
quick thinker with good judgment, but fear is turning me into a sluggish
animal—a bear sliding into hibernation.

"Wind, please stop blowing," I whisper. A cold blast strong enough
to make me stagger in place is the answer.

Wishful thinking is my problem. The promise of buttered seafood
seduced me into believing the wind and sea wouldn't turn foul until
late in the evening, the storm would come more from the northeast,
and this lobster-peppered harbor would remain flat. In reality, the

fifty-four-degree cold front textures my flesh with goosebumps and shoves the boat toward ruin. The sky and ocean froth in a matching Soviet color palette. I don't know what to do.

If idyllic conditions had stuck around, I'd be in a bikini, smearing spit into a mask, scooting bare feet into clown-sized flippers, biting down on a snorkel, and—with one large step—plunging from the deck into liquid turquoise. My skin would practically sigh in delight at the caress of the water. With easy kicks I'd glide toward the Technicolor reef where layers of alien life forms enjoy existences beyond the effects of gravity. Peeping from under a coral head, ruddy antennae pointing toward opposite poles would shift subtly in the current.

Black eyes the size of lead pellets would ask me, "Are you the one?"

"Yes, I am the one. Thank you for your kindness."

My hand would slide down the steel shaft of the Hawaiian Sling spear, putting tension on the rubber strap looping my wrist. The lobster would let the barbed tip hover inches from her belly. Letting go of the shaft, the point would spring forward, crack her casing, and immobilize the animal—despite hard strokes from her powerful tail. Worried the commotion and blood might draw attention from sharks, I'd lift the lobster above the surface and swim one-handed to the boat.

There would be no fresher lobster; there would be no more memorable New Year's celebration. (It would be many months before Dave and I learn the waters around Tilloo Cay are protected; much to my chagrin lobster hunting isn't even allowed here.)

I drag my thoughts to the present and let go of my craving for things to be different. The gloom is increasing. There is little over an hour before dark. The ship's deck vaults with the waves and I am bending my knees to absorb the tilts. My stomach registers the onset of queasiness. I am sweating, despite the chill. A thought of regret bubbles to the surface: I wish we stayed put in the safe harbor of Hope Town.

Under normal conditions, when my husband Dave and I come to an anchorage, the person at the helm turns the boat into the wind and slows the engine so the other can lower ground tackle from the bow. The wind on the nose brings us to a stop, then gently floats the vessel in reverse. The forward mate releases 75 to 100 feet of chain for a long, flat-angled pull. We've done the anchor drill at least a hundred times, burying anchor blades into sand. It is a delicious sensation: the abrupt tug of the boat catching firm to Earth. The sudden rootedness in the midst of flowing sea and air means we're safe and can rest.

This time, however, there is no respite. The chain connected to the first anchor shimmies, an indication the tackle is skidding across a hard sea floor, failing to catch. An aggressive mob of waves and wind gusts and waves push the boat backward. I had wanted to drive forward, retrieve the anchor and chain Dave paid out, and gain more distance from land. But when I rotated the throttle from neutral to full, the machine ignored its controls. Catching on, Dave dropped the second anchor from the prow to keep from losing more ground. Now, pressed by foul seas, it too appears to slither across the ocean floor. The sound of waves colliding onto shore is closer in my ears. Wedged between the force of wind and a wall of land, I feel the press of a crushing execution.

We can't drive out of this spot without an engine, and we can't sail into the wind. Shallow waters frame the exit, so tacking through the direction of the gale is impossible. The only thing I can think to do in this emergency is call for help. The Bahamas Air Sea Rescue Association— BASRA—aids distressed sailors. Sea-hearty volunteers stand by to protect life and property, which they do more than 500 times a year. Someone there will tow us from the menacing shore to penetrable sand and safety.

"What do you think about radioing BASRA?" I ask. "Is it time to ask for help?"

"Well," Dave says slowly, "we're not able to drive ourselves out of this mess and we're not staying put. Yeah, I think times like this are why BASRA exists."

"It's an emergency?"

"Yes," his certainty increasing. "Our property and our lives are definitely threatened."

"Well, OK then."

My thumb squeezes the activation button on the VHF radio. "BASRA, BASRA, BASRA, this is sailing vessel *Wild Hair, Wild Hair, Wild Hair*." Silence. I try again. And again. But no one in a twelve-mile radius hears my call.

This isn't possible. Within hailing distance there are at least six populated islands with harbors shielding as many as a thousand boats. There are three towns with restaurants, resorts, dive shops, and vacation apartments. Nearly everyone in the Abaco Islands communicates with VHF radios. Even if only a fraction are tuned into the common hailing channel, several hundred souls must be listening. BASRA should answer. If they don't, someone else should respond and tell me how to reach the rescue team. People at sea butt into each other's goings-on all the time. The lifestyle is dangerous and sailors cannot afford privacy or pride. This culture of meddling saves lives.

Abandoning the VHF, I drop through the companionway into the teak-paneled salon to lay hands on nautical charts with the phone number of BASRA Headquarters. The office is far away—we're one hundred and forty nautical miles from Nassau—but I'm confident they can contact a local provider, mobilize help. The phone rings too long. Bile coats my throat. The next big gust shudders the ship's canvas covers above deck and the entire boat rattles. I resist the obvious fact: it is dusk on New Year's Eve and no Bahamian volunteer is standing by.

Kelly is not far away—the woman who taught us how to hunt lobsters last week. Dear, sweet, wonderful Kelly—a stranger who motored alongside our boat in her dinghy one morning and announced, "I'm going to get lobster and you want to learn how. Am I right? Let's go." It's a lot to ask someone to move their ship from a safe harbor during a storm just to lend a hand, but she and her mate will do it. They must. I phone.

"Oh, yeah, I remember yoooooou," she slurs. "You should talk with M-mike. He's your man." Muffled grunts and the sound of things crashing sail through the earpiece, drowning hope.

"What's up?" Mike says. I jab the phone into Dave's hand.

"Uh, yeah, hi, Mike," Dave begins with forced nonchalance as he paces the galley. "We are sitting near you off Tilloo Cay. We lost connection between our throttle and engine and we've got two anchors in the water, neither of which is holding. We're backing into shore and we could really use some assistance."

"Ah, broken throttle cable," Mike muses. "That's an easy one. Just replace the cable and you'll be good to go." Mike hangs up. Dave blinks.

"Asshole," my man mutters.

I'm stunned, hollowed out, void of ideas, a plan, or optimism. I have no relevant knowledge or experience to inform what I should do in this moment, despite years of reading about sailing, certificates from a slew of sailing courses, and owning several inland lake boats. What I do have in the length of my body is fear and it pulses evenly. I notice the beginnings of a piercing headache.

"I guess that's that," I say. My tongue is dry and speaking is an effort. "Our cast is complete for tonight's lifesaving drill: husband, wife, damaged engine, slipping anchors, nightfall storm, and ocean."

"Don't forget the rocks." Dave's smile is thin, his pupils small.

"No chance of that." Hot tears of frustration well in my eyes but do not spill. I take a long breath, close my eyes, and on the exhale try to will myself to welcome this new, terrifying reality. This is what you do when you're a practicing Buddhist. It helps.

～～～

I wasn't always a Buddhist, and I'm not sure I fit the title completely. What I practice isn't really religion. Growing up, my parents were pretty typical Midwestern suburbanites in the 1960s and 70s. We all attended church on holidays and other times when the spirit moved us (or when my mom insisted). My only exposure to Eastern practices was on TV; for a period, my brother and I were avid fans of a weekly yoga program. On Saturday afternoons a supple woman with a long ponytail and a tie-dyed elastic outfit would coach the two of us with a soft voice through muscle-splitting contortions. My brother and I would sweat, laugh, and fall over a lot. The big payoff was the ending—the guided relaxation body scan while lying in the grimly named corpse pose. But that was yoga. I had no exposure to Buddhism per se as a youngster and when tidbits about it started landing on my radar in young adulthood, it sounded exotic, way too paradoxical, and irrelevant to my ordinary goings on.

Dave and I raised our kids in the Unitarian Universalist Church. But by the time I turned forty-two, my life was full of inconsistencies. Difficult relationships confounded and frustrated me. I wasn't sure how to address the puzzles within puzzles at work. Our teenagers, too, were walking, talking enigmas. Frankly, I no longer had confidence that I had any idea what I was doing. We talked about it and Dave said he had similar sensations. He and I felt like we had done everything society said we were supposed to do and still something was off. Then one day, while standing in a bookshop, Dave held up a small publication entitled, *Comfortable with Uncertainty* by Tibetan Buddhist Pema Chodron. I mouthed to him across the quiet room, "You so need that!"

It wasn't him who got bit by the enlightenment bug living inside those pages—it was me. Just months later, I was on a worn carpeted floor in a furniture-free Lutheran conference hall with several hundred strangers. A small gaggle of Vietnamese monks glided noiselessly into the room and one settled on the stage. I figured this was the great Zen Master Thich Nhat Hanh, or Thay, as his students call him (pronounced "tie" and meaning "grandfather teacher").

During that first retreat in 2003, the monks and nuns of Thich Nhat Hanh's Plum Village Community introduced me to a secular version of ancient Buddhist teaching. None of the instruction had to be taken on faith. We were encouraged to listen, to test the ideas and practices against our own experience, and to keep what worked, discarding the rest. What was offered floated in and through me like smoke through a forest—practical how-tos and ahas reached into places in my heart petrified by the acids of life. My heart-stone melted and washed clean and left in me something softer and more elastic. The experience sparked an understanding within me of a more expansive reality, the broader context in which my life was set. Doors of healing creaked open. Fresh compassion toward myself and others took root.

Throughout the retreat, the still point at the center of the melting was my breath. I learned how to meditate—how to focus on my breathing while sitting in a calm and relaxed posture. Breathing in, monks told me to pin my awareness to the sensations of my in-breath. Breathing out, I was to keep my mind from wandering by focusing on the sensation of my out-breath. In—aware, aware—out—aware, aware. Repeat and keep repeating for twenty minutes or more.

"OK," I thought, "I'm good at stuff like this. It should be simple. Breathing in, I feel the air moving through my nostrils. Breathing out, I feel warm air moving across my upper lip. Is that the breath coming out? Yes, that's it. In . . . feeling . . . feeling . . . out . . . feeling . . . feeling . . . . Am I meditating now? Something's got to happen soon with all these people here. Is something happening? No, but boy, I can actually hear myself thinking. Hellooo, self! Oh wait. Pay attention to your breath, Heather! You're messing this up. Is this meditating? When is it going to start?"

For the first time, I heard my incessant inner voice for what it was: a selfish, neurotic wreck. Watching my wild imagination peel in 10,000 directions during meditation became more entertaining than anything I'd seen on TV. Eventually, it dawned on me: this unregulated thinking brain with its nonstop, tyrannical voice was the very thing shaping my feelings, opinions, and actions, and I'd been pushed around by this unrestrained bully my whole life. Thay and the monastics showed me that I had a choice. In mindful awareness I could glimpse my true self, the being greater than my small thoughts, the one listening to the prattle who had the power to take a closer look at reality and choose more constructive and loving ways to respond.

This is where I cue the fireworks and strike up the band. This realization completely changed my life.

Here, off the shore of Tilloo Cay, I am fully present to the reality of my fear in the midst of peril. I'm bearing witness to the ways fear is dulling my wits and senses. In the moment, I'm not sure what to do with this insight other than to be with it.

~~~

Dave and I climb outside into the ship's cockpit. Waves splash seawater into the gaps of my clothing, shocking my skin. The clouds are a deeper hue now, bruise-colored. They diffuse the sun's waning rays. Wind blasts my face at about thirty knots.

"Never anchor on a lee shore," I remember being cautioned in a sailing class years ago by a seasoned sailing instructor with a white beard turned yellow by nicotine. "There's too much fetch. The backside of an island, the shore you see when you point into the wind, is the place to find smooth water—even in a gale. If you're on a windward shore and you drift on your anchor, you blow away from land to the safety of open water." Because the wind didn't shift tonight, we're on a lee shore. A half mile away I see air surges mounding saltwater into peaks. The waves gain momentum and height as they careen our way to slam full force upon the vessel. It's difficult for Dave and me to stand. Now I get it. Yellow Beard really meant to never anchor on a lee shore.

Clouds slice evenly overhead, undisturbed by the havoc at the earth's surface. The clouds don't care about the mess down here. In staccato bursts of energy, the wind beats more than water into motion—my hair, the ship's rigging, and palm fronds whip and tangle. The wind doesn't seem to care either. I want desperately to plead my case before the elements, to get righteous and blame someone or something, to beg for mercy. But I am silenced before I begin by nature's disinterest. Forces larger than me are responding to forces larger than them. Something about this deeply offends my sense of justice and I feel my fists clench.

I check the view behind. Limestone spears, carved by time from compressed deposits of an unfathomable quantity of marine skeletons, gleam wet, shine silver in the waning light. A smell of moist stone twists in the salted air. The rock is less than a hundred yards away, and the sharp points seem eager to puncture my hull at the water line, crush our sailing dreams, and send me home to the Midwest in a heap of sorrow and embarrassment.

"All right," Dave offers. "Using the dinghy, let's drive the third anchor away from the boat as far into the wind as we can. Maybe this one will dig in." Good idea.

I bolt aft to unearth our last anchor from the bottom of the ship's storage locker. Lying my belly upon the soaked deck, I reach below to yank out snorkel gear, a diver-alert float, six spare life jackets, wetsuits, cleaning supplies, cans of teak varnish, a long-handled mop, quarts of oil, seventy-five feet of garden hose, one hundred feet of electrical cable, and other matter essential to a life atop the ocean. Of course, the heaviest item—the third anchor—is on the bottom. I reposition my body and lower myself into the hold. The miniature room is dank, slick, and reeks of mold, but it is precious garage space for our floating home. The last time I cleared these contents was months ago. Dave had infuriated me one morning. As an alternative to killing him, I stopped what I was doing, declared the locker space "My Office," and retreated inside until my rage settled. That anger seems to me now a luxury of the truly unburdened.

Boaters from the United States, Europe, South Africa, and elsewhere populate Bahamian anchorages. Monohulls, catamarans, trawlers, and sport-fishing vessels cluster into common hidey-holes—those with protection against the elements and good anchor-gripping bottoms. Crews become friendly neighbors—swapping stories and tricks for a comfortable life in the ocean wilderness. Cruisers spend their days swimming in gin-clear water, exploring the ways of local people, searching for good Internet connections to maintain relations with friends and family back home, hiking ashore, or chipping away at endless boat repair and maintenance projects. Each morning year-round, hundreds of vessels hoist tackle and switch places like a large, slow-motion game of musical chairs. The fact that we are alone tonight at the northwest tip of Tilloo Cay amplifies the foolishness of my choice.

Landing my focus finally upon the anchor and line, I stare in disbelief.

"Bad news. The tackle is in pieces. We need a shackle to connect the anchor to the end of the rope." Why didn't I put this together before I packed it away?

"Huh," Dave considers. He stares blankly at me for several seconds. "If we have extra shackles they're buried somewhere in the bilge. What else can we use?"

My eyes widen as I remember the ship's manifest project—our list of our supplies and their specific storage locations on board—a task also left undone. I want to rant, give voice to our idiocy. Instead, I close my lids, let the urge to holler pass, and direct my imagination to finding a small, circular, muscular link—something that mimics a snap shackle. If this anchor sets, the storm will drive the twenty-four-thousand-pound vessel against the chain with a force I can't calculate. The missing connection must be crazy strong.

"Got it," I declare as I lunge for the companionway padlock—a marine-grade stainless workhorse impenetrable to intruders. "Oh baby, you're gonna be perfect. You're just what we need." Aligning the loop at the end of the anchor line alongside a link of anchor chain, I try but fail to bind the two items together in the lock. They are too bulky to fit inside the padlock's hook.

"Let me see," urges Dave. Sharp-witted under stress, he passes the anchor chain through the loop at the end of the rope and clips the padlock on the emergent link. When he tries to pull the link out of the loop again, the lock holds like a knot preventing a thread from pulling through a needle.

"Threading a camel through an angel's eye," he shouts through a spit of sea. I swoon at his nonsensical poetry and think him the most marvelous man alive.

~~~

Now, the third anchor rests on the ocean floor. Our future depends on a square of penetrable earth and the strength of a padlock. As Dave steps from the dinghy to the ship's deck, he hollers above the bluster. "Let's take up the first anchor because we know it hasn't caught and we don't need three anchors in the water getting tangled." My body seizes with alarm. I think we should leave every hook do its job like cat's claws in upholstery. And yet, my rational mind remembers anchor lines wrap one another when the boat swings. Tangled, they become nearly impossible to retrieve. Tonight's wind will change direction as the storm blows through. Woven tackle would compound our troubles, perhaps immobilizing our ship at the very instant we need to flee.

"All right," I agree from just his side of the fence. "We'll leave the two in place that have the best chance of snagging the bottom."

"Right. We'll get the anchor, and then how about I go on watch first and see what I can learn about the throttle cable from the manuals while you make dinner?"

"Dinner? Can you eat?" I ask.

"Yes. You know I can eat anytime. And you should too. Then you can take watch while I nap. We'll take turns watching and sleeping until the storm ends or we have to do something different."

"Honey, there's no way I can sleep."

"Then just rest. We've got to eat and try to relax to maintain strength because this mess might go on for days. Right?"

The idea of being helpless and afraid for days on end sends a new shock through my system. I breathe to stay steady and take a seat behind the wheel. There is no arguing with his logic. We have no alternative but to take care of ourselves as we wait this out. Palm leaves clatter and bow to the ground in the howling wind. I nod.

Dave goes below to get the manuals while I sit watch and think through our options. Perhaps, I speculate, we'll have a soft grounding when the five-foot-deep keel brings the boat to a halt on the ocean floor before the transom reaches the limestone points. No. This is unlikely, because the nautical chart indicates a steep drop-off. Crashing waves look as if they have eroded a rocky overhang. I suspect we will not catch a break with a soft grounding.

"All right, I'm back," Dave says. He drops a pile of engine manuals at the helm, unsnaps canvas restraints, and rolls protective windows down from the Bimini cover to cocoon the cockpit from the tempest. I climb below.

Canned cheese ravioli plops into the saucepan, still retaining the container's shape. I squirreled the foodstuff deep into a cupboard for a quick source of calories in a pinch; now, staring at the tacky lump of neon-red kindergarten paste my throat clenches and I feel woefully sorry for myself.

"Ravioli, you're disgusting." I grieve the lobster dinner that never was. The tang of canned tomato wafts from the pan and my throat gags as my gut spasms. I'll feed Dave but the fight-or-flight chemicals jumbling my innards will not mix with this. Instead, I fill a water bottle and brace myself inside the heaving ship to drink. *Wild Hair* leaps up and down in the storm.

~~~

After dark—dressed in long underwear, fleece, waterproof bib overalls, wellington boots, and a rain jacket cinched at the neck and wrists—I sit alone at the helm. Rain arrives in sheets now; the world wails. Long lines rigged to raise and lower the ship's sails bang-bang-bang the mast with the repetitive musicality of a toddler. The ship's spotlight is on my thighs and every fifteen minutes I unzip the cockpit enclosure to shine the beam of light through the commotion to judge the boat's distance to shore, to see if the anchor lost more ground. I want things to stay put; I hunger for evidence that the anchors took root. Between glimpses, I wait. I breathe. I acknowledge my racing heart and sweaty palms and search desperately for inner peace. But darkness makes everything ghoulish and exhaustion distorts my imagination. Fifteen minutes before the end of my shift I am completely unnerved and incapable of judging if we are drifting. I leave the beam switched off and sit with the momentary satisfaction of not knowing. Dave will arrive soon, I reason. He'll have fresh eyes and courage to make things right.

Emerging from the companionway a short time later he asks, "How's it going?"

"I don't know. I stopped looking."

The instant the words tumble past my teeth, I know there is something worse than fear: voluntary ignorance. Dave is speechless. He stops in his tracks and looks at me not with anger but with surprise and confusion. I feel shame. It flames my cheeks. I've let myself down by giving up. How could I shun my obligations? I'm not a quitter. This is my boat, my responsibility, our lives; if I love life then I have got to get hold of this all-encompassing fear and find a way to function.

Resting in bed with eyes closed and Dave on watch, I feel the ship's untamed motion from a new vantage. Sloshing reverberates inside the hull with the acoustics of a giant cello. I scan my body yet again, curious to find out how it feels when I'm deeply frightened: steel-tight shoulders, twitchy fingers, rapid and shallow breaths, queasiness, aching temples, and a tickling unsteadiness in the back of my knees.

"Poor thing," I mumble out loud to myself. "You were raised on land in a field of safety. You do not know what to do in this situation." I wrap my arms around my middle and give my body a hug.

My attention drifts to a conversation some months back. "Have you scared yourself yet?" our boat broker Tom Harney from Jordan Yachts asked me. We crossed paths in Florida months after buying *Wild Hair*.

"No, not really," I remarked. Dave concurred. "We've gotten into some pickles, but we've gotten ourselves out of them," I boast.

"Well then, you haven't been sailing yet. You haven't really been sailing until you've been really scared."

I guess tonight—more than a year into our planned five-year voyage aboard *Wild Hair*—is my first real sail. I laugh. Sailing is horrible.

My mind floats to a story the Buddha once told. A young girl ran from a pair of snarling tigers through the forest. She fled for her life but stopped short at the edge of a steep cliff. There, with the thud of cat's paws approaching from behind, she climbed over the edge to dangle from a broken tree branch. The tigers caught up and paced above her head at the cliff's edge. In an unfortunate turn, three more tigers arrived to circle below her perch; they too were ravenous. The girl could not climb up; she could not drop down. But beside her she saw a single strawberry growing through a crack in the cliff. It was plump, ruby red. So what did she do? She ate the strawberry, free from fear of what would happen next and in full awareness of the beauty and safety of the present moment.

Tonight, the story of tigers rings of spiritual discipline. A determination stirs in my core to experience the richness of life here and now, even in the company of my fear. Yes, I am afraid, but I am other things too. I am safe in this moment. The boat is sound. Danger is present, but the future has yet to be determined. This understanding changes something in my middle; a pressure lightens.

Just then Dave bursts through the door. "Heather, get up. We're moving, now. The wind shifted north and we're not holding."

Above deck, Dave shines the spotlight through the gale. Waves collide into and swirl around immovable stone contours; the undulation tricks my eyes into believing the rocks are breathing predators. I feel stalked.

This is our moment of reckoning, the seconds before we save ourselves from ruin or resign ourselves to shipwreck and loss. Dave studied the engine manual and crafted a plan requiring him to run above and below deck while I steer a precision route. He figured out how— from below— to goose the engine as I throttle forward. But we need someone at the bow to haul anchor lines as we drive to keep the ropes clear of the propeller shaft. I must steer at all times, so he must dash between two jobs.

"OK," he instructs, "we will set the boat speed at a somewhat aggressive pace to power into these waves. The biggest mistake we can make is tangling anchor lines around the propeller. Right?"

My mind leaps to a time months ago when we were anchored in Florida's Intracoastal Waterway off Amelia Island for a picnic and the current spun the boat in a circle. The anchor line tangled the keel and prop shaft, and *Wild Hair* hung at an odd angle in the out-flowing tide. That day, I shifted the engine into drive just as Dave coached, "Keep the boat out of gear."

THUNK. The wrapped line snagged the prop blades and choked the engine to a quick death. Dave jumped into the placid Florida water to cut us free, but once the line gave way, the boat floated loose at great speed; he only made it back on board because I was able to motor toward him. A repeat of that event in this storm would put my man in violent waters and catapult our vessel—with one person on board and no motor—toward the land that we longed to flee.

"Oh my God, no. Let's not wrap the prop."

"As we go forward, time means distance, right? I figure you'll need to drive forward for about twelve seconds on a 265-degree heading before we're on top of the first anchor."

"Yeah, I'm with you."

"When the first anchor is on board, you'll turn to a 320-degree heading and drive toward the second anchor. I'll take up the line as you go. If you maneuver that course, we shouldn't snag lines."

I picture the conditions in my mind: our position, the placement of anchors relative to magnetic north, the length and angle of submerged ropes, the contour of the sea floor. I imagine how the route, boat speed, and timing of turns will play.

"Yes—got it. That should work." As I climb over the seat to my station behind the wheel, I notice my shoulders are rock-hard and my headache continues to pound, but my limbs are swinging more freely than they did. Thoughts are flowing with more ease, too. This body continues to be locked in fear, but now that there is something helpful for me to do, fearlessness is also on the rise. Fear and fearlessness coexist side-by-side in the tight quarters of my interior.

To add a measure of safety, I switch on the floodlights to illuminate the deck—the zone of our rescue operation. Suddenly we are center-stage on a surreal movie set. Fans and rainmaking machines churn off camera. We have one take—one chance to save ourselves and our ship. Live or die, I am fully present in the company of my fear and prepared to end this crisis.

The engine starts when I flip the power switch and push the ignition button. Together, we set the throttle into a position we both agree sits at the right boat speed. Dave goes below and manually adds engine power to the limit of the extended throttle. The engine pitch changes and *Wild Hair* starts moving. I maneuver the boat on the predetermined course through the blow as Dave darts out the companionway toward the bow, shouting, "Here we go." He holds on as the boat pitches, exercising care that his feet keep their grip in pooling rain. I drive twelve seconds—counting time in my head—toward the closest anchor at the 265-degree heading. Dave hauls chain onboard. Just as I reach twelve, he hoists the first anchor.

Perfect. That's one anchor on deck and one more to go. My hands turn the wheel to the next heading. The ship's compass spins in its mount, settling at 320-ish. Dave, flexing arms and legs, spends brute force lifting chain from the sea.

Normally we use our anchor windlass—a motorized system of gears—to haul heavy chain from the ocean into a storage locker beneath the deck. But the windlass is slow and prone to jam. As I watch the scene through slanting, meteoric marbles of rain, my husband's arm and back muscles strain. His vulnerable human frame glows wet and silver in stark contrast to the black backdrop of night. He yanks hard as if his life depends upon his action. It does. I imagine him collapsing at the bow with a heart attack. Oh, my love, this is all too much.

"We're free," Dave finally whoops, barely audible above the blow. "Drive us outta here, baby!"

The glow of floodlights makes me blind to our surroundings, so I steer by video picture, retracing the route of our arrival captured on the GPS monitor. Dave climbs below deck again and—in the engine room—adds more power. I try to relax my fingers, which are clamped frightfully on the wheel. It's difficult to take in our new reality: the ordeal is over. I don't hear the soft popping sounds.

"Oh, look. Happy New Year, sweetie," Dave says as he climbs into the cockpit again and fireworks burst above our starboard rail.

"Happy New Year to you, my love," I reply as I kiss him. "Why did we leave Hope Town?" I ask, realizing what a nice holiday we could have had there, on the safety of a mooring ball in a protected harbor.

"I don't know, but we'll sit down tomorrow and figure out everything we learned to keep this from happening again."

He's right. The analysis can wait. In this moment I want to cherish the simple thrill of being free of danger. Breathing, I try to relax my shoulders, but they are brick hard and stubborn. Finding calm won't be easy. So, I look outside myself to the horizon for the roots of peace. Colorful bursts drizzle over palm trees, lighting up the night. A happy New Year blows in.

REFLECTIONS ON FEAR

The day Dave told me the dollars-and-cents worked for a multi-year sailing sabbatical, I leapt at the chance to go, not only because it was an old dream of ours to sail over the horizon, but because I thought a sailing stint would serve as an advanced environmental training program. Time spent navigating the ocean wilderness would give me a chance to learn directly from nature—the Great Atlantic Teacher. Traveling by wind, I would surely come to know a profound intimacy with natural systems and discover new ways to live lightly and harmonize with nature's order. I eagerly signed up for what I thought would be the spiritual voyage contained within the physical journey. I was ready to be remolded by waves, sand, the sun and moon into a new woman (someone unflappable, with the capacity to see clearly and smile at life's bruising blows). So even at the moment off Tilloo Cay when I was most afraid for my ship and my life, I didn't doubt I was on a path toward something beautiful, good, and true—such was the strength of my desire to learn what Earth herself has to teach.

In retrospect, actually I did something *right* on that fear-filled night by keeping my awareness on the experience of fear in my body—how it ebbed and flowed moment-to-moment. Uncomfortable as they were, the sensations of fear were like blips and dashes of Morse Code saying, "Heather, this is important; you're in danger; watch out; do something." Fear kept me alert and tuned in, but the strength of my fear just about monopolized my inner circuitry and stopped other blips of vital information from flowing. Feeling every sensation as it came up was an effective way to keep the energy and the information moving *through* the circuitry. Had I not monitored the experience attentively, I suspect fear would have short-circuited the works and overwhelmed me completely.

Scanning my body to take in the discomfort of fear seemed a natural thing to do under the circumstances. I had already developed the habit in my role as executive director of a nonprofit environmental justice organization. I was well-practiced at paying attention under moments of stress to the sensations of strong emotions—fear, anger, jealousy, sorrow, or greed—and created little gaps for myself between the stimulus (panic or anger) and the response (striking out); this mindfulness practice offered me a half-second or so to pause and think about my reaction. For example, just the knowledge that a comment really got my goat made it possible for me to think, "Well, that's interesting." Once I stopped, I could take a breath, avoid compounding the difficulty, and keep moving toward my goal. Learning the secret behind interrupting knee-jerk responses before they got me into trouble was a delicious fruit of Buddhist practice, one that helped me sustain a long and effective advocacy career under frequently contentious circumstances. (This simple spiritual technology also helped me be a better mother.)

My life as a change agent began when our son and daughter were babies. In 1996, I launched the nonprofit conservation organization I eventually spent nearly a decade and a half leading. I loved the work and arrived to the office each day with enthusiasm and a sense of purpose. My coworkers and I helped inner-city communities redesign schoolyards, slow and reroute traffic to increase neighborhood walkability and bikeability, create farmers' markets and skateboard parks, and put important landmarks on the national register of historic places. We partnered with local residents and bought toxic land, mitigated contamination, and reengineered nature's ecological services back into the landscape (slowing and absorbing storm water, recharging drinking water aquifers). As a result, vibrant parks and community gardens, clean air, pools of clear water, and neighborhood friendships flourished where blight and relative isolation once stood.

Over time, I noticed the growing concern around climate disruption triggered emotions in me that I didn't allowed myself to fully feel. What started in the 1990s as a small pang of guilt over a life of American privilege became a panicky kind of irritability for *someone* to do something—*anything* to avoid or at least soften global catastrophic loss. Subconsciously, I blamed myself for being too dense to know exactly how to take action. Then I'd remember I was already tied up managing a conservation organization and conclude it was OK I didn't have energy to overhaul my and my family's daily routines. Still, every once in a while, I'd have to shrug off the guilt I felt driving my car from Madison to Milwaukee for a meeting instead of taking the bus. This is how my back and forth emotional responses to climate change sizzled just below consciousness; it was an internal tug of war that was at once quiet and fairly hellish now that I think about it.

When I decided to hand the organization's reins over to a new director, I promised myself I'd bring one question with me as I met the Great Atlantic Teacher: *how am I to live in a suffering world?*

The boat broker's observation that I hadn't really been sailing until I'd been really afraid applies, I think, to climate disruption. Sitting with this in meditation, turning the idea over like a rock, I discover underneath it a partial answer to my central question. I had never really let the reality of climate change *land* in my body. I'd not honored my fear as valuable information. Sure, I knew about climate change because I had a master's degree in environmental science, but I kept the pain of it at a professional, analytical distance, protected myself from the full *emotional* impact of what's coming down the pipe. For years I adhered to a form of voluntary ignorance. The storm off Tilloo Cay taught me ignorance is worse than fear, because not-knowing actually compounds danger.

Not anymore. This is my life, my planet, and caring for it is my responsibility. If I love my life (which I do) then I am going to get hold of this fear, experience it fully, and not let it jam my circuitry.

I suspect many of us in the climate movement have either dipped into the data in an analytical or emotionally distanced way, or jumped in headlong and started reacting like inexperienced horse trainers banging pots and pans to frighten the herd into cooperation. By making ever-increasing amounts of noise and focusing on apocalyptic projections, we have sent a lot of folks running to the hills in denial. Of those who remain, some are in the heated lather of panic and others are nearly catatonic, completely overwhelmed. A precious few are operating somewhere in between: calmly, clear-headed, and with skill in the midst of strong emotions resonating in their bodies.

I wonder what would happen if each of us coexisted with the uncomfortable fear, anger, heartbreak, and guilt reverberating in our bones and chose to behave in ways that are composed, plain-spoken, and compassionate. What if the climate concerned were like horse whisperers who modeled the fearlessness and wise action they wished others would adopt?

The wild sea on New Year's Eve showed me fear is a fire capable of burning away complacency, but like all flames it can overwhelm and destroy. Observing and monitoring the strong emotion of the moment, putting it in check, helps me remain calm, understand the danger I am in, and grow more skillful in my response.

I offer the meditation below as a tool to help myself and others dwell peacefully in the presence of strong emotions. Practice it when sitting comfortably with your body and mind quiet. Take your time. Breathe slowly in a natural and relaxed way as you focus your awareness on the text. If (when) your mind wanders, bring it gently back to the sensation of your breathing and the words of the guided meditation. You may enjoy having a light smile on your lips.

Breathing in, I am aware of the sensation of my in-breath.
Breathing out, I am aware of the sensation of my out-breath.
In, breath sensations.
Out, breath sensations.
(Ten breaths)

Breathing in, I am aware of sensations present throughout my body.
Breathing out, I relax into the sensations throughout my body.
In, body sensations.
Out, relax into body sensations.
(Ten breaths)

Breathing in, I am aware of feelings that are in my body.
Breathing out, I relax into the feelings that are in my body.
In, aware of feelings.
Out, relax into feelings.
(Ten breaths)

Breathing in, I am aware climate change threatens life on Earth.
Breathing out, I feel the sensations of climate change awareness
    land in my body.
In, aware of climate change.
Out, feel climate change reality in my body.
(Ten breaths)

Breathing in, I am aware climate change threatens life on Earth.
Breathing out, I feel climate change awareness stir my emotions.
In, aware of climate change.
Out, feel strong emotions.
(Ten breaths)

Breathing in, I relax into the sensations of
    climate change in my body.
Breathing out, I relax into the emotions climate change stirs.
In, relax into my body
Out, relax into my emotions.
(Ten breaths)

FORT
LAUDERDALE

MIAMI

FLORIDA

*Biscayne
Bay*

EVERGLADES
NATIONAL
PARK

KEY LARGO

*Florida
Bay*

PLANTATION KEY

MARATHON

KEY
WEST

**FT. LAUDERDALE
TO KEY WEST**

2 THE MIDDLE WAY

Ft. Lauderdale, Florida // March 15, 2008

It is Saturday afternoon and our son and daughter tip the taxi driver and lift colossal suitcases from the trunk. Where on *Wild Hair* will we put those, I wonder? Their northern clothes look hot against their pale skin. Sweat flattens their heads. I take in Maggie's most recent purplish hair shade. It's not bad. Eland looks the way he always does when he comes back from college—as if he hasn't slept in days.

Descending the gangway, the two exchange a private joke with sibling intimacy. A burst of Latin music erupts from a car at the beachfront nearby. Our children flash big smiles at us, and I swell with delight that—after months of being apart—they are here from separate universities for a spring break sail to Key West. My stomach flutters with excitement for the week ahead.

And yet a knot of tension in the middle of my back pecks at my awareness. Dave and I at sea are as savvy as Keystone Cops in a canoe using bicycle horns for paddles. I'm convinced something will go wrong. I want to give my children the vacation of a lifetime, but am not sure I've figured out how to do that yet.

Dave and I warned our kids that in these first twenty-three days of sailing *Wild Hair*, misfortune struck with the frequency of commercial breaks in prime time television. It's humbling. *Wild Hair*'s keel struck bottom eight times. Three times the ship spun in the current and slammed into concrete piers, bending posts and streaking the hull with black smudges. Once, we swung too wide in the current while docking and drove our bow into the stern of the boat in the next slip; luckily, both vessels were unscarred. Five times *Wild Hair*'s engine overheated before someone successfully diagnosed the cause. Twice tow boats came to our rescue and pulled us to mechanics on shore. We dragged the boat's anchor, strayed off course, grew seasick, and—on an overnight sail under a moonless sky—hallucinated an attack of seals.

Things broke. We regaled our children with stories of Dave's repairs: the ship's autopilot, lower bilge sensor, upper bilge float switch, stuffing box, and refrigeration compressor. They heard about the handsome young mechanic in Charleston who—while repairing *Wild Hair*'s engine only ten days into our boat ownership—offered encouraging words.

"Don't y'all worry none. This happens to everyone. It'll take ya about two years."

"Two years before everything is fixed?" I shuddered to think of the costs and delays. "Why no! Two years b'fore y'all know how to fix everything y'self!"

We told Maggie and Eland about sailing day number thirteen when, in St. Augustine, the engine's kill switch didn't work. At the end of a very long day, we couldn't shut the motor off. We tried everything, but the beast thrummed. Desperate, Dave and I disconnected the fuel injectors and finally the machine sputtered to a deafening silence just after ten at night. The next day, another mechanic pointed to the breaker on the electrical panel responsible for powering the kill switch; the electrical connection was mistakenly shut off. The mechanic flipped the toggle, which repowered the connection, and handed us a bill for $127.

We talked openly with our children because we didn't want them to think our journey was without risk or that their vacation would be stress-free. Also, we hoped they would arrive ready to pitch in when things went haywire. Now, with minds open to any possibility and calibrated toward adventure, Eland and Maggie drag roller bags down the marina gangway. Their eyes slide left and right, taking in the view.

Wild Hair is floating in a congregation of mega-yachts. I watch their expressions as the splendor of inconceivable wealth seduces their imaginations. When the pair clears the four-story private cruise ship docked to our port, they lay eyes for the first time on our family's sloop.

"Oh, I thought it would be . . . bigger," says Eland as he delivers a distracted kiss to my cheek.

"Yeah," says Maggie as we embrace with sticky arms. She steps back to look at *Wild Hair*. "This is pretty small compared to the others. Is this it? I mean . . . it's cool and all, but it's not like those!" She ogles what must be a thirty-million-dollar boat next door.

"You had us going, guys," Eland laughs, giving his dad a manly back slap. "I thought you really had something!"

Ouch.

Knowing how much they would love to drive off in any one of the yachts nearby, I laugh. Those ships carry jet skis, deep sea fishing boats, and glossy, classic runabouts in trunks five-times the size of our ship; they would comfortably accommodate all my children's friends and their friend's friends. The mega-yachts ooze elegance from their chandeliers, rare wood, and the brass buttons adorning their young, tan, and muscular crew. This marina—like others in south Florida—represents a staggering concentration of wealth. Our little boat is the poor relative of these glamor queens.

How can I express the fondness I feel for our modest mass of fifteen-year-old fiberglass, steel, and cloth? *Wild Hair* isn't a head-turner, but her modest size and strength means freedom. She's strong enough for Dave and me to sail by ourselves to any port in the world. Our canvas sails reflect more environmentally-friendly sensibilities—big motor yachts burn tens of thousands of dollars of diesel in a single trip, while we fly on renewable energy: wind. We manufacture electricity for our lights, refrigeration, and electronics not with a diesel-burning motor but with a wind generator and solar panels. I am completely satisfied by our simple craft and want for nothing more.

This is the boat Dave and I imagined before we thought about kids. While dating, we ignored "keep out" signs and strolled hand-in-hand through marina gates toward boats that beckoned. Neither of us had a lick of experience. But as we walked mazes of piers, gabbed about the boats we admired, guessed at the purpose of different rigging, and declared destinations we'd someday aim our ship toward, we fell in love with each other and an unconventional future. We knew someday we would stop whatever we were doing, and give each other the gift of our time and undivided attention.

Standing with our children, I try to see *Wild Hair* through their eyes as seagulls beg for scraps—*cree, cree*—from the fishing boat across the channel. This sloop is the reason mom and dad sold their beloved childhood home in the Midwest. Perhaps the boat doesn't look worthy of that sacrifice. Or maybe our stories make them feel vulnerable to seasickness, calamity, or disaster. Alternatively, it's possible their parents' happiness may ignite a lifelong passion in their hearts for sailing. Perhaps the fact mom and dad—at the peak of their careers—stepped away from jobs they loved and a lifestyle of comfort to make real a youthful dream demonstrates that the freedom to explore new approaches to life can be more rewarding than more of the same business as usual. Or maybe the decision to set sail will sharpen our children's resolve to accomplish their own schemes, giving them courage to thrive. Who knows what sense they'll make of this in time.

～～～

Waking up the next morning, we wedge shoes and books into storage lockers and prepare to depart. We scratch the day's path onto a paper chart, fill the engine's overflow reservoir with coolant, and shift binoculars and winch handles to the helm. Perfect Paul—the automated voice on the VHF radio— forecasts an idyllic day. I drove last, so it's Dave's turn to drive and he motors us from the slip and toward the mouth of Port Everglades.

I look and it's where I expect it to be. The uniquely curved, mint green high-rise sits at the edge of the beach like a twenty-two-story cereal box. Confirmation of its continued existence makes me giddy. I start at the pool deck and count eighteen stories up and three balconies over, but the action of the sea and the sameness of repeated patterns confuse me. Somewhere in the mix of upper story condos is the former home of my long-deceased grandparents.

Here was magic. This was a place where a young girl found freedom from teachers and annoying assignments, leaky winter boots, and wool coats incapable of stopping Chicago's wind. This was the destination of airplane rides at a time when I wore my Sunday best to travel: a straw hat, fishnet stockings, white patent leather Mary Janes and a matching pocket book. Here, Christmas was truly green. Grandma's cooking was strangely Italian, to satisfy the appetite of her new husband. Restaurants had jazz bands and seafood. Streets levered skyward for boats to pass, creating miles of congestion. On one of these balconies my grandmother tucked my brother and me in at night with a ticklish game of "mousy-mousy" and we slept on reclining lounge chairs under the moon, wrapped in moist air, while listening to the primal heartbeat of the Atlantic's waves. Right here is the locus of my deep affection for Florida.

My left hand grips one of the port shrouds and I lean past the deck toward the building.

Wild Hair's bow parts and curls water beneath me. The sea burbles.

Life is full of unimaginable loops. That young child never imagined she'd be the object of her own intrigue: a sailor taking her vessel out to sea.

"Can you count to the eighteenth floor and find the third balcony from the east? That's grandma's mother's home. Remember, I told you . . ."

"Wow—that's neat!" says Eland.

"There it is!" says Maggie.

But my message is lost in the newness of their total experience. My eyes alone linger toward the height. To them, it must look sun-bleached and aged like the other beachfront high-rises. The marble floors of the mezzanine ballroom and the unpretentious chandelier the size of an inverted water tower are invisible. There is no way to convey the richness of my 1960s experience, one woven with parental orders for perfect manners and the depths of a grandmother's generous love. They were extraordinary times. I want to tell my children about their ancestors, how they lived, what they hoped for (as far as a granddaughter can understand), but this is not the moment; the kids are too distracted. Suddenly I miss my grandparents and wish with an aching heart that they were standing on their balcony, watching, waving a scarf at me the way they waved at passengers aboard the QE2 ocean liner.

Dave navigates past rip-rap at the harbor's entrance and turns south. An 85-degree breeze flutters my t-shirt, cools my belly. The boat bounces playfully in two- and three-foot seas. Medicated patches behind everyone's ears keep seasickness at bay. I beckon the kids to leave the cockpit, scramble forward, and find handholds near the mast.

"Watch how I raise the sail and you can do it next time. You're going to want to see what I'm doing with the lines, but keep looking up and checking what's happening with the sail. That's the focus of attention."

As *Wild Hair* vaults, I grab the smooth, blue line sprouting from the base of the mast, loop the loose end twice around the gut-high winch, plant my feet firmly on the deck, and pull hand over hand to raise the canvas. Short puffs accompany my rhythmic pulls. After many tugs, the line won't come. I look straight up, arching my back and squinting into late morning sun. The narrow top corner of the triangle is about a quarter of the way up the aluminum pole, but things are snagged. The trailing edge of a batten—a stiff bar stitched into the sail to give it shape— caught the rigging.

"The wind is swirly," I say. "Dad is trying to feather the boat straight into it." My hands ease the line a matter of inches until the batten relaxes. "We'll wait for the boat to wag in the wind so the batten swings free." An out of sync wave, which is made by a speeding vessel with a towering fly bridge, bounces the three of us off balance. We giggle at the silliness of our stumbles.

"Mom, are we going to play Schluge later?" Maggie asks.

"You bet. And, your Dad and I are going to win this time." The kids moan to contest my claim. "It's true. Your dad and I are going to win the coffee mugs. I bought some beauties."

"Maggie . . . " Eland drops his chin, tilts his head slightly, and sends Maggie a look of supreme seriousness—one reinforced with a waggle of his hand indicating they alone have the invincible brain-wave connection for success. "Those mugs are ours."

He's probably right. When the four of us play Schlutenglaggen—a complex card game similar to canasta that our family embellishes with an evolving made-up language—the duo wins ninety percent of the time. Beyond competitive, our children are telepathic.

The sail slackens and rattles; the batten swings to the centerline. In a burst, I haul in long pulls, but soon my strength gives way. The sail is shy of the top, yet the weight of cloth suspended is too much for my limbs. So I wrap the line a third time around the winch, snug it in place, and insert a foot-long handle into the drum.

"Crank this," I say to Maggie. She forces the winch handle away from her body and down, then toward her middle and up. This single rotation lifts the mainsail a few more inches.

"You know, mom, you can't beat Eland and me at cards, because he raised me." Her arms crank. Eland sniggers.

"Oh yeah . . . that's right. Your brother raised you like you were a lost child and he was a kindly wolf that took you into his den. I remember you telling me."

"Well, he did."

"Okay, Maggie. Is that why you're now studying to be a zookeeper? Are you trying to understand your brother?" We laugh. "When you do, explain him to me."

One turn at a time, she hauls the giant canvas to its height six stories above the sea. When everything is taut, I cleat the halyard and give Dave the okay. He adjusts the course.

Ka-whap! The sail snaps like a colossal bed sheet as it fills with air. The boat angles fifteen degrees and accelerates in a rush. I love the sensation of launching; my breath arrests every time in a flush of exhilaration.

"This is so cool," says Eland as we work our way from handhold to handhold to the cockpit. "I'm glad I packed two novels by Patrick O'Brian. Our boat is just like those in the Napoleonic Wars. I mean, it's different, but it's really the same in its size and the way it moves. Don't you think?"

"Absolutely," says Dave from behind the wheel. "We just need cannons to blast the other ships from the water."

"Yeah," says Eland. "Maybe we can conquer an entire fleet!" A trawler cuts in front of our path.

"Dad can be the captain but I'll be the admiral," I chime.

Wild Hair is working her charms, drawing out the simple joy of being together. Or maybe it's not the boat but time in nature relaxing us and making differences in age, interests, and experience insignificant. Each of us slowing down, letting go of the pressures of land, and relishing the abundance of laughter and attention between us. The change is incremental but sure. At some point as we exited the port, we crossed into a happy place where we are free to be completely ourselves, together.

Tonight we will travel as far as Miami and anchor in Biscayne Bay, beyond the cruise ships. We will swim, play cards, and slurp bowls full of spaghetti. Tomorrow we will dinghy to a state park, put our feet into the sand of a beach rated by a tourism magazine as one of the best in the country, hike to the top of a musty lighthouse, and drop anchor near Key Largo at Pumpkin Key.

Plans won't unravel until day three of our trip.

~~~

I stir Monday morning from the deep mattress of the master stateroom to the click-click-click of the stove's pilot light and the opening and closing of galley cupboards.

"I want tea to calm my stomach," says Maggie. She looks green despite the quiet anchorage.

"Do you want me to make you some pancakes?

"Maybe later," she says without conviction. Balancing a full mug of ginger tea in her left hand, she hikes over her brother (drooling on the pull-out bed in the salon) and steps into the forward stateroom without spilling or waking him. I look at the chaos of what is normally a tidy ship. Teak walls and floors glow. Comfy leather sofas the color of antique china encircle a table with an embedded gold compass rose, but socks and yesterday's clothes litter the table and seats. Silk throw pillows stitched with orange coral branches are squashed by suitcases. The flat-screen tv and bookshelves lining the walls port and starboard under panoramic windows are still visible, but the shades are drawn so the room is dim. At the chart table on the portside, an impressive cluster of radios and electronics look business-like and ready for the day, but there are a pile of cellphones charging.

My hand reaches for the coffee pot. I'm starboard, in the chef's galley equipped with a gimbaled, stainless three-burner range, rows of cabinets, a separate refrigerator and freezer, a double sink, and ten linear feet of spice bottles in handcrafted teak racks. A breath of air meanders through a port light, swirls Eland's frame and delivers to my nose the aroma of a college male in need of a shower. How I miss this boy.

Retching pierces the quiet. Maggie—visible only from the waist down—is standing on her mattress, and through the hatch overhead, throwing up outside on the deck. At once I realize she's not seasick; the girl must have the flu because she's sniveled and dwindled since she arrived. The ensuing clatter startles the boys. There is commotion. Then, as families do, we press on.

We are taking the Atlantic Intracoastal Waterway (ICW) to Key West, making headway through Biscayne Bay, Card Sound, Barnes Sound, and Jewfish Creek. The ICW is a protected route first envisioned by government leaders in 1824—a time when lumber, minerals, harvests, and people moved easiest atop rivers. Officials were wary of the Atlantic Ocean's dangers—pirates, armies, and storms—and they imagined the alternative course to bring the residents of a young nation together. To begin, they charted the coast's natural inlets, salt water rivers, bays, and sounds, and then plotted canals across land masses. It took 116 years to carve and dredge the watery highway, but light-draft commercial, military, and recreational vessels incapable of traveling the distances between major Atlantic ports came to depend on the 1,200-mile lane from Norfolk, Virginia to Key West, Florida. A 2006 study by the Florida Inland Navigation District determined the ICW is now responsible for eighteen billion dollars of economic output in this state alone.

While the ICW can be beautiful, I prefer the freedom of the open ocean. The Atlantic is a straighter shot and more adventurous. Our boat is rugged enough to manage the ocean's volatility, but with the kids onboard, Dave and I thought it wise to be conservative. On the inland route, the barrier islands dampen the strength of winds by a third. The twelve-foot-deep channel and narrowness of the winding corridor tames the rolling waves. What's more, in this stretch, the creeks and rivers meander through stretches of stunning and unspoiled territory.

Late in the morning, a squadron of pelicans assembles in the air, circles, and suddenly peels apart as members of the regiment dive headlong into the drink.

"Look," I say, standing, "there must be a school of fish near the bow." The animals' plunging doubles as slapstick humor. Their antics are as graceful as a demolition derby.

"You'd think they'd give themselves concussions," Eland says. Then a pod of dolphins surfaces off the port rail and darts into the action; fins stitch above and below at breakneck speed. At once, the creek and air above churn with mammals, birds, and fish. Agitated water reverberates between the islands. It is a raucous fresh seafood buffet. But this meal, it is really more than eating, isn't it? I'm seeing pelicans bounce from the backs of dolphins and the slippery noses of mammals tag and toss birds eight feet overhead. Dolphin and pelican are cavorting playfully, sharing a meal, and fully enjoying each other's company. It is a kind of laughter, an inter-species frolic.

"Are you seeing this?" I ask.

"This is so weird," says Eland on top of Dave's grunt of disbelief.

"Wow," says Maggie. "They're like . . . " she searches for the proper term, "hunting cooperatively."

"And they're fooling around to boot," I add. A pelican jumps from one dolphin's head to another and then leaps high above the turmoil before it spins in the air and plummets into a dive.

"Oh man . . . did you see that guy?" Eland laughs.

"That's hysterical," says Dave. "He must be really hungry." Another pelican flies across our path six feet above the water; a fish tale flaps in his beak. Two more drop from the sky bomber-style just past our reach.

Many minutes of happy feasting pass before the predators are full, and disperse again to their hidden ecological niches. The air and water settles, and I'm left wondering if the peculiar aberration was real.

"That was so neat. I want to see it again," I shout loudly to dolphin and pelicans.

"Yeah, but they ate their fill, mom. They don't need more," Maggie reasons. "That's not the way nature works." She wraps herself deeply into her sweatshirt, leans on a cushion, and closes her eyes. My daughter is reminding me nature isn't here for my entertainment. Of course I know this, but I am eager to meet the untamed world in a new way: in person. I spent almost all of the past two decades cooped up in an office, learning about and caring for nature, but most of that work was done behind a computer monitor. I'm excited to get outside and witness flora and fauna in their natural state, see them cavort and thrive.

I turn in my seat and look back, but the fish, pelicans, and dolphins are gone. "Hey everybody," I call aloud, "where did the party go?" Greenish-brown water ripples in the breeze as the boat's engine drones. What just happened was strange. I haven't a clue how to extrapolate meaning from the feeding frenzy. A scalloped cloud moves in front of the sun, and the world falls into shade. This was one of several spontaneous encounters with wild things during our brief ownership of *Wild Hair*. They always seem to follow the same story arc: rising suddenly and fading quickly, leaving me startled and mixed up.

I realize that it's not entertainment but understanding I seek. Was the eating rush just one more example of survival of the fittest in a dog-eat-dog world? Perhaps, but that characterization doesn't capture the joy and playfulness I witnessed. I saw a balancing of life and death with profound equanimity in the exchange between giving and taking. And although it was spectacular to us, the animals merely came together, and moved on. It was a commonplace occurrence that happened so fast I didn't have time to take it all in before the creatures got their fill and quit.

The predators stopped consuming when they could have had more—a stark contrast with how modern humans generally behave. Perhaps the universe was offering me a lesson in moderation. In the natural world, creatures consume only what they need to stay strong and vital, and leave the rest so others may thrive and multiply. At this seafood buffet, no one filled their pockets with crackers or asked for to-go bags. They ate what they needed and stopped, putting their trust in the natural continuation of plenty in order to survive.

These animals are fearless. Now that I think about it, only a few species stockpile food to survive harsh seasonal conditions. But full-blown hoarding and gluttony are nonexistent in nature. No other species on Earth accumulates wealth like humans; resources continually flow in and through natural systems so life can thrive.

It would be a profound transformation if civilization suddenly stopped exploiting nature and people, suddenly halted all efforts to stockpile wealth and resources. How crazy would it be for individuals to put something away for retirement or college, but only what they need and no more? Giving away excess wealth in the prime of life—passing it on in amounts greater than even the most generous charitable donations—is counter-culture. And what is "excess wealth" exactly? There is a common understanding of "excessive displays of wealth," but I've never been privy to a serious conversation about "excessive wealth" in my life. The idea reeks of anti-capitalism. Only spiritual leaders appear to be extreme enough in their values and beliefs to trust the universe for the continuation of plenty, people like Christ, Buddha, Gandhi, and today's monks and nuns. Who knew Thich Nhat Hanh had so much in common with pelicans?

It's odd that when I start thinking about the distribution of wealth, my mind leaps to money. I started down this path by thinking about the circulation of wealth in nature. True wealth is really not about money but about what it buys: life-supporting resources, access to food, water, shelter, security, leisure, and personal freedom. Out here, dollar bills and coins have no inherent relationship with wealth. Pelicans and dolphins stopped eating when they had enough, and their moderation became a nonmonetary form of generosity to the fish. And by not over-hunting the fish, they ensured the continued health of the ecosystem. There's no room for greed in the equation. Moderation—not taking too much so others may live—is the generosity of sustainable relationships.

Earth is speaking loud and clear: moderation is key to answering my question about how to live in a suffering world. The Buddha heard Nature's lesson 2,600 years ago: he called a moderate life The Middle Way and he urged his students to do their very best to avoid any and all extremes. For the Buddha, there was no reason to be overly austere but neither was there reason to overly indulge in pleasures and comfort. We should neither ignore nor wallow in emotions. As we move through the day, we'd be ill advised to spend too much time on any one thing, be it eating, drinking, talking, sitting, or sleeping. Living the path of the Middle Way, he said, will unlock for us a treasure-trove of joy.

Suddenly, I remember my first retreat with Thay, I wanted to participate in an initiation ceremony of sorts in which I made public my commitment to study and practice Mindfulness. I had to write a brief statement of aspiration, so I made an impressive laundry list of everything I planned on doing to make the world a better place by applying my new skills. (Frankly, I wanted the monks and nuns to know I meant business.) At the conclusion of the ceremony, each of us were awarded new Dharma Names—a thoughtfully picked moniker to remind us of our truer, more expansive selves. My Dharma Name: Non-Attainment of the Heart. What?—I thought. Didn't they read my statement and everything I plan on doing? Oh . . . I guess they did.

This was my first lesson in living the Middle Way.

~~~~~

At midday, the ICW grows shallow—too shallow. We bump and plow the keel through Blackwater Sound into Dusenbury Creek before we come hard aground near red channel marker "48A." Charts and navigation instruments tell us we're in the center of the dredged channel, but the boat is stuck in place. My scalp prickles with worry.

"How deep is it supposed to be here," asks Eland.

"Twelve feet," says Dave. "Something is off."

"Are you in the channel?" I ask, climbing over sweatshirts and binoculars to look for myself at the ship's GPS.

"Yeah," Dave confirms. "We're where we should be as far as I can tell."

"Is the tide going down?" asks Maggie between blows of her nose.

"Hmmm, no. We're approaching high tide. And it shouldn't matter because the charts give us what's called 'mean low water depths.'" I say. "The dredged channel should be at least twelve feet at low tide. We've heard it might be shallow in spots, that dredging isn't happening like it should, but this is nuts."

Dave powers the engine in forward and reverse; *Wild Hair* rocks but doesn't budge. "Well," says Eland, "what do we do?"

Dave and Eland are suddenly energized; they concoct an elaborate plan to hike their bodies to the end of the boom, swing the boom sideways over the water, and use their weight to lever the keel from bottom as I drive the boat toward deeper water. Maggie goes below to make chicken bouillon.

"Here, dad, gimme a leg up." says Eland. Dave laces his fingers atop his knee for Eland to stand. But between the boat's hip-hops, Dave's staggering, and the boom's wiggling, Eland can't find a stable purchase. The men regroup.

"Hold on, I got this," Eland says as he steps atop a winch and makes to climb onto the Bimini canvas.

"Hold on, dude," I call. "That canvas won't support your weight. It'll tear."

Near the mast, Dave says, "We can get on the boom from here."

"Oh, yeah. Great idea," Eland strides to his dad's side. Dave explains safe places for our son to put his feet to scale the mast. Just then, a fishing boat with two, two-hundred horsepower outboards rumbles to a halt alongside our hull.

"You need a pull?" asks a barrel-chested man as a smaller fellow and a pair of preteens look on.

"Why, yes," I say, "that'd be great!" I release one of *Wild Hair*'s spare lines, wrap it around a stern cleat, and toss the loose end to the helpers.

"Where did all the water go?" I ask.

"Don't know," says the local. "But don't trust your charts. These waters will do-in a deep draft sailboat." Fishermen cleat the free end of the line to their stern rail, drive an about-face, and tug. *Wild Hair* leans but stays put. They increase horsepower and heave again.

Without warning, the wheel at the helm spins uncontrolled.

"Wait, wait!" shouts Dave. "Stop!" But the fellows are deaf from distance, wind, and engine. Our sailboat stretches to the side and, after a long moment, gives way to float free. Dave mutters in panic.

"What's going on?" I ask.

"We should be pulled going forward, not reverse. Did you see the wheel?"

"What's wrong, dad?" Maggie asks as she returns to the cockpit.

"The rudder hit bottom and dug in" Dave says. "When they pulled the boat, the force was on the rudder—it twisted as it held. That's why the wheel spun." Eland steps toward the stern rail as our helpers whoosh to *Wild Hair*'s side.

"Is that a problem?" Maggie asks.

"Pins can bend or the rudder can snap when we go backward because it's unprotected in that direction. The skeg post protecting the rudder is *in front* of the blade. We'll have to see if we have steerage as we get going."

One of the boys tosses the end of our line to Eland and shouts goodbyes and good lucks.

"Hey, thanks, man," Eland shouts. "We really appreciate it."

"Yeah," I holler. "We'll tow you next time!"

I climb behind the wheel and gently motor forward—giving "48A" wider clearance.

"Give the wheel a spin," says Dave. I swing the wheel right and left; the stainless steel feels cool to my heat-swollen fingers. The boat S-turns in response. "Any looseness in the feel of it?"

"Nope. She feels same as always." I exhale tension, relieved that the rudder steers true. "Bruised but not broken. We got lucky."

At a slower speed and with more attention to water depths, Dave and I take turns pecking through Buttonwood Sound, Cross Bank, and Cow Pen Cut. Channel markers indicate the whereabouts of the deepest water, but our keel scrapes and drags bottom a dozen more times without grinding to a halt. Progress is stressful. Finally, in Cotton Key Basin about a quarter mile off Plantation Key's shore, we call it quits.

The basin's shallowness dictates we anchor in the middle of the bowl in seven feet of water. What's unsettling is that, at high tide, the depth should read more than nine feet. With only seven feet of water, *Wild Hair* will sit on the bottom when tide slips away.

We drop anchor. Fifty feet of chain jumps in after it, lines on the deck loop into neat piles, and the engine kill-switch flips. I put my hands on my waist, pull my elbows behind me, and raise my chin to unlock the day's tensions. The sky shimmers flamingo pink and orange as the sun dips below the horizon.

~~~

Dave wakes me in the night. "Heather!" he whispers. A wave of alarm shoots up my spine.

"What?"

"If we get stuck, really stuck, then whoever comes to our rescue can claim they are salvaging a wreck. They have legal rights to take our boat."

"What? Not our boat. Not with us on it. That makes no sense." I take a moment to consider his reasoning. "No. I don't agree. Go back to sleep."

"Yes, I don't know the details of when a normal tow turns into a salvage operation, but at some point International Maritime Law kicks in. You know that."

"No one can have our boat because we just got it! Let the bastards come and try." I take a breath, roll over and bend my pillow to a new shape. Sea swells gently rock the mattress. "This is just a crazy middle-of-the-night worry. Let's talk about it tomorrow."

"Heather, it's true," Dave says. "And we can't drive south on the ICW any further because for some reason there's a lot less water than there should be. Charts show the waterway getting shallower ahead. Using tide, we could travel to Key West like we planned if depths were as charted. But they're not."

My eyes open and I spy my dive watch glowing at the side of the bed:

03:47. There's no sleeping until this gets resolved. "Yeah, I saw that, too. Things will get worse."

"I don't want to go back the way we came. We barely made it this far. If we turn around we might get stuck, need a salvage, and lose our boat."

"I'm not agreeing with you on that. I won't let that happen."

"Neither of us knows for sure what we're talking about. That's why we have to be careful. We're floating now. Tomorrow I want to call a tow company and have them get us out of here using their local knowledge. If they get us grounded it's not a salvage operation; it's a tow gone wrong and it's on them."

"That makes sense, but what about the kids? They were looking forward to seeing Key West; they'll be so disappointed."

"It doesn't matter. We'll still have the boat and we'll be able to give it another try next spring break."

Of course.

~~~~~

Dave calls the tow company after breakfast while the kids and I ready the boat for departure. Outside, the sun floods a clear sky, but strong gusts swirl. *Wild Hair* jumps in place.

"You can either guide us out of this hole before we're stuck or you can tow us after we're aground," Dave jokes with the tow captain. "It's up to you ... OK ....OK .... Yeah, you're right .... OK, will do. That's a plan .... Yeah, bye."

"Well?" The kids and I stare.

"They're forecasting a storm. He'll get us to the ocean through Snake Creek, a skinny path between barrier islands. But the tow captain isn't pulling us until high tide Thursday morning—after the storm dies."

"What?" The three of us blurt.

"High tide will give us the best chance for success and *Wild Hair* will be easier to tow when the blow dies."

"We're sitting in Cotton Key Basin for three days?" My question hangs in the air. Eland and Maggie shoot each other a look. The anchor chain creaks. Turning to Eland and Maggie, I say gently, "There goes our chance to visit Key West, kids. I'm sorry about that but that's just the way it is. Pretty much the rest of spring break will be right here, cozied up in the storm."

"Yeah, yeah, that's cool," says Eland.

"By the time we're free, we will have to turn around and point north so you can catch your planes," Dave adds.

"But the good news is we've got lots of food, games, and movies. What do you say, can we still have fun?" With hands on the kids' forearms, I give them a little shake to unseat discontent.

"No problem," says Eland with a chuckle. "I've got my books."

"I'm cool, mom. I can be sick anywhere. It might as well be here." Maggie smiles.

As they remove their shoes and retrieve their books, I scan the chart to find Snake Creek, double check the plan. Dave pours himself the last cup of coffee.

"Should I make another pot?"

"Oh, yeah," I say, looking forward to my next sip.

I draw a finger across the paper to trace a noodle-thin route between islands. The movement stops on the number "3." Snake Creek is labeled only three feet deep. "How is this going to work? It looks to me like the makings of a wreck."

Dave reads the route and shrugs. "It's our route. They know our draft and said they'd get us through."

"Ughhh—we're putting our trust in their local knowledge. I hope they're not recent transplants from New York." I stand on my toes and look out the salon window toward shore.

The surface of Cotton Key Basin is textured with two-foot-tall white-tipped peaks; the water is dusty gray, but the sky remains flawless and bright. Toward the east, land and vegetation stretch flat on the horizon. Snake Creek, the promised break in the span, is invisible. Shifting my weight onto my heels, I think this trip is a family legend in the works: the vacation spent gazing toward land beyond reach, while nursing the flu without medicine, and worrying over a scheduled shipwreck.

The blow builds through the morning and *Wild Hair* begins to hobby-horse: bow up, stern up, bow up, stern up. Walking below decks is difficult, so we grab handholds to avoid falling. I scan the chart again, this time hunting for better shelter in the storm. Plantation Key Marina is nearby; it's far into the shallows, but they might have a dredged channel. I phone.

"I see you out there. What's your draft?" asks the marina staffer.

"We're five feet."

"Oh no, there is no way you can get in here. Years past, yes, but you can't come in here now."

"Are you sure?"

"I'm real sure. I've got a measuring stick in the water. Now-a-days, our high tide is fourteen inches below what used to be the typical low water mark."

"What? High is fourteen inches *below* what's supposed to be low? Why is that?"

"Don't know; it just is. So you're safest where you're at. Stay put, now."

Sitting at *Wild Hair*'s small navigation booth, I put my elbows on the desk and drop my face into my hands. A dozen thoughts spin. As a sailor, I know the tide floods and ebbs twice each day, raising and lowering water levels at this location by nearly two feet. If high tide is that far below low tide, then three feet of water is missing across the expanse of Cotton Key Basin, and maybe the whole of Florida Bay— the territory between the southern tip of Florida and the keys. An absence of this much water is big news. It makes no sense when sea level is rising; melting ice and expanding seas have already added eight inches to the ocean depths. Why is the water level in the Bay falling? And what's with the laissez-faire attitude of local fishermen and marina owners—people who depend upon the water for recreation and income. Why doesn't this set off alarms in their heads?

Dave and the kids are lost in separate literary plots, so I venture online to untangle the mystery of missing water in south Florida.

I learn the state goes through predictable cycles of dry spells ending in hurricane floods about every twenty years. Eager to support agriculture and development in the twentieth century, the federal government went hog-wild and built 1,400 miles of canals and levees, hundreds of storm water pumping stations, and straightened the meandering Kissimmee River. Experts decided the water levels in Lake Okeechobee, the Caloosahatchee River, St. Lucy Estuary, Lake Worth Lagoon, Biscayne Bay, Florida Bay, and the Everglades weren't to their liking, so they altered nature there as well. Eventually, the network of dikes, dams, and drains sent 170 billion—that's billion with a "b"—gallons of fresh water into the Atlantic and Gulf of Mexico each and every day. Sucked dry, the state could now grow sugarcane and sell homes to northerners like my grandparents seeking warmth. The downside was that the Everglades—one of the most ecologically complex fresh and saltwater wetland systems in the world—nearly evaporated. Compounding the problem, Florida kept flooding periodically, even after the state had removed the land's ability to store fresh water during cycles of drought.

Finally, a state commission circled the wagons, and a 1995 study under Governor Chiles's leadership stunned everyone with its evaluations: if the state's ecosystems continued to fail, Florida's tourism and commercial interests would tank, and health hazards would skyrocket in the form of rising urban crime and shortages in drinking water. Five years later, the Comprehensive Everglades Restoration Plan came off the press and got rolling. The plan prescribes—over the next 30 years—increasingly elaborate engineered solutions that would cost billions.

Here *Wild Hair* floats in Cotton Key Basin—a brackish fresh- and saltwater pool in the Florida Bay missing nearly three feet of water. We're in the eighth year of a historic drought—the worst to strike America since the Depression. The trifling rainfall that does fall on the state is—by design—ushered summarily to the ocean. The itty-bitty capillaries linking the Atlantic to Florida Bay—passageways like Snake Creek—are far too insignificant to revascularize Florida's life-giving organ—the Everglades. The latest engineering schemes are pricier than expected, stakeholders war over turf, politicians vacillate, people complain of a stink around the accuracy of scientific calculations, and funding for engineered solutions is stuck high and dry in Congress.

Wild Hair's rigging whistles in a fresh current of air. I rearrange my legs in the snug booth, slide bare feet across the varnished teak floor. Maggie snores lightly with a book on her chest. My shoulders droop.

South Florida's tangled mess of good intentions and conflicting agendas is a familiar, troublesome acquaintance of mine. For too many years as an environmental justice advocate, I was the lead goose in an advancing V-formation. I stuck my neck out and honked noisily, flew as straight and fast as I could, carved a slipstream for others to follow, and took the inevitable fire. Usually, the shots were glancing blows, ordinary occurrences in the machinations of consensus-building. Occasionally, they were personal hits that stunned. A few times, blows made me somersault and nearly fall. But every strike of competing agendas, derailed deals, political vacillations, media challenges, and budgetary hits took something from me—ample slices of energy, peace, optimism, and heart. Mostly I responded by denying my vulnerability and ignoring my pain because the people and landscapes I served were worse off than me. In my thirties, I needed medication for a bad stomach. In my forties, I needed pills for an irregular heartbeat.

When I set sail on *Wild Hair*, I was a battle-weary, broken-hearted activist in desperate need of rest. Sometimes waves of grief and guilt overwhelm me while sailing in a kind of post-traumatic stress. Thoughts turn negative. I tell myself I haven't done enough because there is so much more suffering in the world than when I started. I should have accomplished more and I blame myself for unskillful actions. Sitting in front of the computer, I take a breath and realize I've stepped away from the job, but I left nothing of my concern behind.

Advocating for ecological and social well-being is frustrating, stressful, and only occasionally successful work. No doubt advocates working in south Florida ask themselves the question I hold in my heart: How am I to live in a suffering world?

Scientists warn of a lag time between the release of carbon dioxide and its effects. (A 2015 study by the National Academy of Sciences reports there is already enough greenhouse gas in the atmosphere to "lock in" south Florida's soggy fate.[1]) Florida's wetland landscape and its extensive coastline, where 75 percent of its population live, is about to be flooded by rising sea level. Human habitation is unsustainable and literally groundless. This makes Florida's historic growth that benefited early investors like my grandparents nothing more than a hundred-year-old Ponzi scheme. Later investors are sure to lose out eventually.

No one knows how many years or decades will pass before south Florida dips below the surface. But salt water will inevitably sluice backward into last century's channels and ditches, inundate neighborhoods, overwhelm drinking water aquifers and roads, poison inland agricultural lands, and prompt a mass climate migration of residents—from Miami to Tampa—out of the state to higher ground in America's interior.

I can picture the day when Cotton Key Basin is no longer shallow from freshwater diversion and drought, but is overtaken by the swelling sea. Floridians will need help and our nation will want to provide compassionate aid as in times past. But when mass migration away from the rising sea begins here, people will also be fleeing the broiling desert of the southwest, the water parched stretches of the Great Plains, and the desiccated landscapes of Mexico, Central America, and many Caribbean island nations. Populations will migrate from Africa and Australia, and flooding fields in Southeast Asia.

We are on the brink of a global climate migration.

The storm knocks *Wild Hair* at an angle. I lean back and look at my children. Smart, healthy, curious, and caring, the two dwell on the shore of incalculable possibility. We share this time in history, but their generation will live though the full onset of climate disruption and bear witness to society's response—the beauty and the ugliness. By the time they are my age, they may know civilization's fate.

"The storm is here," I say to the room.

"Yeah," says Maggie, stirring. "But we've got everything we need for now. And we've got a plan."

"That's right," says Eland. "Hey, let's play cards. I can read later. Let's do something together, because this time is special."

"Great idea," says Dave.

"Why, yes it is sweetheart," I say. "Our time is very special and I want to enjoy every minute of it with you."

REFLECTIONS ON THE MIDDLE WAY

"I don't know how you get Americans to understand natural resources are finite," said my Caribbean colleague. "It's easier for me. I can see the end of my island; people here know the limits of what we must care for."

The speaker was Crafton J. Isaac—assistant biologist at the Ministry of Agriculture, Forestry, and Fisheries on the island of Grenada in the West Indies. It was October of 2011, and Dave and I had sailed *Wild Hair* all the way down the Caribbean island chain. I sat in the mint-green painted office on a seat too low for the table feeling awkward and a bit silly. The room was above St. George's Fish Market, near the center of town, so our conversation was punctuated by the racket of fish trading: ice cubes cascading into plastic bins, trucks revving their engines without mufflers, women vendors shouting creole words. My nose tingled with the sweet-salty smell of fresh-caught snapper. Just being in this place with Isaac—a local resident *and* a MacArthur Fellow—was an adventure.

"Look—people in the islands live simply," he said. "They don't consume much and their lifestyles emit very little greenhouse gas. Islanders have no extreme behaviors to moderate in order to curb rising sea level, but they will suffer from infrastructure loss and more and worse hurricanes, nonetheless." My pen scribbled across my notebook, the writing barely legible.

"This is an island of finite financial and natural resources, so if Grenada is left to pay for climate change adaptations, our resources will become rapidly depleted." Isaac went on to describe how the social fabric of his small nation would tear as the environment degraded and people jumped ship. Throughout the conversation, the worry he carried for his homeland was palpable.

"We will lose our resources to the same countries who make the warming; our fish are moving north. Grenada will become a diaspora or an ecological and social disaster—like Haiti." Then, he surprised me when he said, "The fate of Grenada hinges on America's middle class and its ability to wake up to the fact that they too are victims of exploitative development and consumerism. Grenada needs Americans to stop consuming so much. The standard of living on TV is not desirable or achievable. Their behavior is unsustainable for everyone on the planet."

Our talk that day makes me think moderating actions in south Florida could produce worldwide benefits. If Americans on the front line of sea level rise act with skill, they will help people and governments around the world blunt the effects of climate disruption.

Unfortunately, things are worse off now in south Florida than they were when Dave and I traveled with the kids. The National Climate Assessment named Miami one of the most vulnerable cities to the effects of rising sea level.[2] Already, people wade through streets and first floor lobbies at high tide. Saltwater spices drinking reserves in aquifers up and down the coast, forcing elected officials to close wells and scrounge for water inland. Even so, construction cranes still pepper the Miami skyline. At the time of this writing, real estate in Miami Beach is selling at a dizzying sixteen hundred dollars a square foot. What's really worrisome is that developers remain free to disregard the science of sea level rise because hardly anyone talks about it: Governor Rick Scott—ensnared by the mirage of unending expansion—reportedly banned state employees from even saying the words "climate change" or "global warming." So, I guess I can't be too surprised by Florida residents who—when surveyed—said they worry little or not at all about rising sea level and other effects of warming.[3]

Sometimes, as I breathe in meditation, I see that what pushes us off the path of the Middle Way. The troublemakers are the ancient challenges of the Three Poisons of the Mind—greed, hatred, and delusion. I know greed, hatred, and delusion to be sneaky, insidious phenomena—weeds with fragile leaves and robust root systems making them difficult to extract. During Maggie and Eland's spring break voyage, for instance, I had a greedy craving to give our kids a vacation so enjoyable it would wipe away the residual guilt I carried over selling our home. My desire for pleasure was strong enough to delude me into believing we could travel the ICW to Key West, despite the low waters. When Dave suggested others might take ownership of our boat in a salvage situation, I responded with hatred toward those who might try. Traveling south through Florida Bay, my greed, delusion, and hatred might have put the family on a path leading to nothing but distress.

Our globally connected economies and cultures magnify the effects of the Three Poisons. Individual indulgences that once appeared to be benign expressions of greed, hatred, or delusion have spread and been adopted by people around the world. Our habits, taken collectively, have the potential to become catastrophic. For example, I didn't realize as I sailed the Atlantic how harmful my greedy craving for meat was to the planet (especially my consumption of beef and dairy), but eventually I learned that animal agriculture is a leading producer of greenhouse gases, both CO_2 and methane.[4] A main cause of the planet's warming is that animal agriculture is expanding to feed society's collective demand.

Leaders in Miami are caught by the Three Poisons, too. Phillip Levine, the mayor of Miami, clings to a form of delusion—the scientifically unsound belief that the city will be viable in five hundred years—and he aims to spend a half-billion dollars to elevate every road in the city and plum the landscape with a system of giant pumps.[5] Local architects advocate to turn the city into a floating island—a Venice of the United States.[6] But if the whole of South Florida is awash, can this dot on the map survive—physically and financially—as a lone floating city in an increasingly tumultuous sea?

Florida International University professor Henry Briceño doesn't think so. Briceño publicly questions the viability of any long-term development scheme for South Florida.[7] His reasoning: the state's historic shoreline once stretched as far north as Lake Okeechobee and that's where it's likely to return. Briceño asserts that it's reasonable to keep the city dry and economically viable for the short term, but sooner or later people must move. Briceño articulates a moderate approach, the Middle Way.

I'm no climate scientist, but I am a spiritual ecologist formally ordained into Buddhist leadership, with an advanced degree in environmental studies who spent a career helping local units of government and city residents figure out how to better get along with Nature. To my mind, a realistic, center-of-the-road coastal adaptation plan for South Florida—one that neither ignores reality nor hyper-reacts to changing conditions—should include a well thought-out migration and relocation plan. Relocating millions is an epic task, and the worst flooding projections are only ten to thirty years off. Now is the time to figure out where and how in America's interior new business investments, jobs, food, water, shelter, roads, schools, health care facilities, and other public services will come to life. It's on this generation to invent creative and entrepreneurial responses to climate migration as a reasonable and wise expression of the Middle Way.

～～～

Thich Nhat Hanh says it best when he observes the Middle Way (personally consuming less land, energy, material goods, and food) is an antidote capable of waking us up and bringing the hungry expansionist view to an end. When we walk the path of the Middle Way, we naturally remove the poisons of greed, hatred, and delusion from everyday life.

The path of the Middle Way becomes personal when we observe greed, hatred, and delusion at work in our *own* consciousness. Then, we develop the know-how to recognize and diffuse them in society, too. To train myself in the art of neutralizing the three poisons and taking the path of the the Middle Way, I wrote and practice this guided meditation.

Breathing in, I see myself contributing to the well-being of others.
Breathing out, I smile to generosity.
In, contributing to well-being.
Out, smiling.
(Ten breaths)

Breathing in, I see the poison of greed in me.
Breathing out, I see ways to moderate greed.
In, see greed.
Out moderate greed.
(Ten breaths)

Breathing in, I see the poison of hatred in me.
Breathing out, I see ways to moderate hatred.
In, see hatred.
Out, moderate hatred.
(Ten breaths)

Breathing in, I see the poison of ignorance, denial, or delusion in me.
Breathing out, I see ways to moderate ignorance, denial, or delusion.
In, see delusion.
Out, moderate delusion.
(Ten breaths)

Breathing in, I aspire to moderate my lifestyle.
Breathing out, I aspire to step on the path of The Middle Way.
In, moderating lifestyle.
Out, choosing The Middle Way.
(Ten breaths)

WEST PALM
BEACH

WEST END

OLD BAHAMA
HARBOR

GRAND BAHAMA

**FT. LAUDERDALE TO
OLD BAHAMA BAY,
GRAND BAHAMA**

FORT
LAUDERDALE

3 SKILL

My two feet step on ladder rungs, my hands grab teak rails, and my muscles flex as I rise from the galley into the cockpit. The surrounding marina—buildings, piers, bushes, cranes—is tomb-dark. A transformer hums nearby. Tires on a lone car drum across the seams of the bridge nearly overhead. I can tell my body wants to sleep because my limbs hang with fatigue and my blood refuses to flow. Thoughts creak. Everything feels off, and yet I am thrilled to be awake—like a child defying sleep at a slumber party. Tonight's journey begins at 01:00 to arrive at the nearest Bahamian island before sunset. We are pointing *Wild Hair* straight east, into the deep, to a place we've never been. We are testing new skills on a trip that promises to be as carefree as a tax audit.

Stretching my back, I notice the air. I always notice the air these days because my skin relishes the tropical moisture. Sweet, spicy scents of nameless things growing and dying at the edge of land and sea fill my lungs. In stark contrast to the winter air currents of my Wisconsin homeland, this gentle breeze compels me to relax my defenses. I find it seductive, like the sparkling eyes and sly smile of a Southern gentleman. My body opens; senses intensify. My nervousness about the trip loosens as I melt into the balmy air.

"Are you ready?" Dave asks, bounding up the companionway. His own jitters electrify the night's shadows.

"Huh . . . We'll see."

We are virgins tonight, inexperienced sailors making our first Gulf Stream crossing. Caution is warranted. We're cruising into the Bermuda Triangle and a lot can go wrong. Rogue waves and mini-cyclones—surgically precise fingers of God—can appear from nowhere to sink ships. The vessel might lose a through-hull connection and take on water faster than we can pump and bail. Or we may collide into a shipping container bobbing just below the surface, one spilled from a freighter in a storm. A hole in the boat would sink us in moments. Alternatively, I could misstep and fall overboard; Dave, unawares, might glide away and leave me to watch *Wild Hair*'s sails, awash in starlight, shrink on the horizon. Eventually, I'd have no choice but to gulp in a salty, liquid grave. Even if he were to see me fall into rolling seas and quickly maneuver for rescue, Dave could easily lose sight of my head, especially at night. These real worries are why, in the last few weeks, we fixed a life raft to *Wild Hair*'s transom and installed an extra pair of high-capacity pumps. We wear life jackets when sailing offshore and physically attach ourselves to the boat, clipping and unclipping as we move about the deck.

Tonight we will meet the Gulf Stream, part of the North Atlantic Gyre, one of five twists of ocean current and wind stirring the earth's surface. This Gyre flows west from Africa, splits at the bulge of South America, and moves north into the Caribbean Sea. The strait between Florida and the Bahamas acts as a flume: the broad flow narrows to a mere sixty-two miles and the water's concentrated energy accelerates to 5.5 miles per hour. Tonight's challenge is to sail across the Gulf Stream, to bisect the flume.

The term "Gulf Stream" refers to the part of the North Atlantic Gyre that begins at southernmost Florida and tracks north along the East Coast of the United States and Newfoundland. A river in the ocean, the stream creates weather; it carries warmth and saltwater from the equator to Europe. Astronauts can see the flow's thermal gradients with the naked eye from space.

I am standing barefoot on *Wild Hair*'s deck in a sleeping marina—a boatyard located miles west of the ocean in the center of urban Fort Lauderdale. I scrunch my toes in the dew and imagine myself kin to early explorers. Juan Ponce de León was the first captain to log his encounter with the Gulf Stream. In 1513, he sailed his south-pointed ship backward for a day because the current was a stronger force than the favorable winds. Years later, Spanish captains used the corridor like a conveyor belt at the airport to speed north. Benjamin Franklin, benefiting from conversations with seafaring captains in both the US and England, was the first to chart the enormity of the current's influence. It was he who coined the name Gulf Stream.

In modern times, the current's energy potential lures environmental researchers to its depths. Scientists envision underwater turbines spinning as water moves, generating sustainable power. Others look to the thermal energy potential of the temperature differential between colder deep and warmer surface waters.

At the point where Dave and I will cross, water will move under our hull at thirty-million cubic meters per second. It's a massive force. I've plotted and replotted a course to take in the influence of this phenomenon. It's tricky. I mapped our departure and arrival points and estimated—given the boat's speed, time in the current, and sideways velocity of water—the compass heading we must take to arrive on target. Sailboat captains have a more difficult time navigating the stream than power boaters because our ships move more slowly, spend more time in the current, drift farther off course. Because of this, I'm as certain of my course as a drunk walking a white line.

How will liquid Earth feel when the solidity of land within swimming distance disappears? Where will my thoughts stray when my eyes see only water? Popular guidebooks say I'll know I am in the Gulf Stream when I see changes in bubble patterns on the surface, differences in color as ocean temperatures warm, and an abundance of wildlife— from whales to turtles—catching rides on the conveyor from southern habitats. All that is fine, but I fear a violent ride.

The predominant wind, blowing from north to south, lifts the north-flowing stream into steep, breaking waves. Like petting a cat in the wrong direction, the surface bristles. In the Gulf Stream, an everyday wind can stack waves tall enough to sink a ship. If *Wild Hair* is struck broadside by a large curl, she may roll over and down.

We've done a year's worth of daytime and overnight coastal sailing, but nothing like this. Our multi-year sabbatical started when we bought the boat in Wilmington, North Carolina, bopped down the coast, and eventually cruised into the harbor at Key West (without the kids). This past summer, we kept *Wild Hair* at a marina in St. Simons Island, Georgia—north of the hurricane line established by our insurance company. Weeks ago, we sailed south again and Dave and I have been sitting in the marina-boatyard in Fort Lauderdale—the ideal launch pad for a journey to the Bahamas. It's one thing to skim down the coast, and quite another to point the bow straight into the great unknown and disappear over the horizon.

Aware of the dangers, I monitored weather forecasts for the past several weeks, looking for a window of safe conditions. I listened to crackling broadcasts from the National Oceanic and Atmospheric Administration (NOAA) detailing wind speed and direction along with wave height and sea swell. Day after day, inhospitable cold fronts consumed the region and forecasters issued small craft advisory warnings. Fellow boaters said wait, be patient, don't venture into the Gulf Stream in anything but ideal conditions. But I feel pressure to get going.

Our plan is to rendezvous with family for a Christmas celebration in the islands. Eland, Maggie, and my brother and parents will share a historic captain's home on the beach of Elbow Cay. Dave and I will anchor inside the enclosed harbor of Hope Town, and dinghy to shore to join the family in beach walking, exploring the village, and swimming in the resort's pool. Everyone will come aboard *Wild Hair* for day sails and island hops.

All this is good, but there's an old sailing axiom: true mariners don't sail on a schedule, because so much lies beyond our control. Oops. We have Christmas as a deadline. Maybe this is why family members are placing bets on whether or not the kids will celebrate their first parent-free holiday.

The ideal weather pattern for traversing the Gulf Stream is two days of gentle, southerly winds (one day for seas to calm and one day to travel). But this is December, and northerly cold fronts are screwing with our plans. At this point, I'll settle for a minimum of fourteen calm-ish hours with moderate winds from anywhere but north. That should be enough time to journey from our present location, nine miles up Ft. Lauderdale's New River, to West End, Grand Bahamas Island. The problem is fourteen quiet hours have eluded us until now.

At 01:00, the wind off the coast will be easterly at less than ten knots. Waves should be two feet. We cannot sail east, directly into the breeze, but we can motor our way across. On the other side, the weather will come around and be perfect for scooting the Bahama Banks for the three days it will take to get to Hope Town. But the change in weather will be lousy for any more crossings until sometime after Christmas. This moment is the only window for a safe Gulf Stream crossing before Christmas.

Given this, I'm surprised we are the only people awake; I thought others would be departing tonight for their Bahamian holiday. We're the only ones awake and moving about. The realization we are alone unsettles my gut. Somewhere in the distance, a train lets loose a lonesome cry.

Our solitude also means there is nobody to help release dock lines as we depart. Our vessel's directional control in reverse—like the backward control of any sailboat—is supremely limited, especially in a current. On top of that, *Wild Hair* is squeezed as tight as an egg in a crate between mega-yachts. The most difficult part of the voyage may be leaving the slip.

"What's the direction and speed of the New River current?" Dave asks as he drops a corner of paper towel into the water. I watch the inch of glowing white drift from mid-ship toward the bow.

"It's flowing out," I say, "At maybe . . . three knots?"

"Yeah, that sounds about right."

Together we try and work out how various forces will push the boat once the ropes let go of pilings: the water is moving in one direction, a breeze nudges opposite, and the spin of *Wild Hair*'s propeller drags our tail left when we back up. We size up obstacles in our path and think through the best defensive sequence for releasing dock lines. I imagine collisions-to-be and reposition fenders. Dave double-checks the engine fluid levels and belt tensions, and then arms the engine start at the circuit panel below. Sitting at the ship's wheel, I flip the switch to power the engine starter near the helm. Locusts chirp in the night. The ship's flag flutters. Scanning the cockpit, I confirm necessary equipment is handy: binoculars, paper charts, winch handles, life vests, rain gear, a flood light, water bottle, log book, flashlight, and pen. Everything is ready. It's time to go.

I press the electric engine-start button. Silence. I expected the same deep-throated vroom that bellowed just hours ago when we started the engine to confirm the refrigeration was charged and working. But now, there's no "*wannawannawanna*" sound of a tired battery struggling to catch, no click of the starter solenoid doing its job. There is only screaming quiet as I push repeatedly on the start button. My mouth hangs agape.

"Well?" Dave asks from the galley through the companionway opening. My left hand takes up the flashlight to illuminate the action of my right, the futility of my effort.

"Shit!" He does an about face toward the bookshelf of repair manuals. I join him in the search below decks, scour trouble-shooting journals, and look for relevant pages.

"Here it is," he says. "I think we have corrosion on the starter solenoid that's interrupting the electrical flow from the push button to the engine. Why it picked now to lose connection, I haven't a clue."

My partner opens the engine room doors and slithers down the bulkhead to an uncomfortable half-sitting, half-kneeling crouch. I pass him tools from a bank of drawers. He fumbles low in the bilge.

"Yeah," Dave confirms. "Salt water got in here somehow; or maybe it was just salty air. It's heavily corroded. Let me have that snarly metal file, would you?" I pass him the instrument and he shifts position to get a better angle, grunting. In rhythm with the metal scratching he complains, "I retired from surgery so I didn't have to operate in the middle of the night!"

My sweet husband. A marriage, raising two children, and managing separate careers left us unprepared for the intensity of cohabitation in forty-five feet of living space, 24/7, for months on end. His steady kindness and support nourished our children and me for decades. But time on a sailboat makes plain stark differences in our characters.

He was captain of an operating room for twenty years. Now Dave hands out orders with the ease of a waitress schooling a fry cook. Sometimes the frequent, unconscious telling-me-what-to-do peeves me. I remind him our sailing resumes are the same. I ran a business and have my own ideas. There are multiple ways of doing things and we approach challenges differently.

The way I understand it, Dave tends toward a *reductionist* method of problem solving: he takes a thing apart to understand the component pieces, how the bits works. I see *emergence* at work and accept that a thing cannot be understood by dissection alone, because elements interact in unexpected ways. As unrelated things combine, creation bursts forth with something altogether different than the parts—like when two atoms of hydrogen combine with one atom of oxygen to make water. Water? As an ancient Sufi teaching says: you think you understand that two and two makes four, but you understand nothing until you understand "and."

Maybe the lens of optimist and pessimist is at play, too. After all, it is said the ideal sailing team has one of each—the optimist gets the boat off the pier and the pessimist gets it safely back. But tonight, he gets no argument from me. We team up on the solenoid's repair because Christmas hangs in the balance.

After giving the connectors a rough polish and smearing sticky electrician's grease on the works, Dave reconnects wires and uncurls his body. I scramble to the helm. This time, the starter brings the engine to life.

The navigation equipment surrounding the wheel glows. The eight-inch GPS monitor asks if I agree to the terms and conditions of its use and I agree to them all, whatever they are. The VHF radio roars static so my left hand adjusts the squelch as my right hand peels back the stainless dome guarding the compass. The compass disk floats horizontally in its crystal ball of oil, the dial shining red to protect the skipper's night vision. I pry the cover from the autopilot display and my eyes jerk in a double-take. The unit, which should be glowing green, is dark. I cringe.

"Dave, the autopilot's dead!" Yesterday the two of us reinstalled the newly varnished cockpit table onto the helm's grab bar; in the process, we must have guillotined delicate wires feeding the autopilot. This is a blow. Similar to a car's cruise control (but maintaining direction rather than speed) the auto pilot lightens the strain of driving. Sailors consider the autopilot an extra crewmember. Now, ours is dead. Hand steering for ten weeks in the Bahamas will be inconvenient and tiring, but it is not a justification to stay. Besides, old-school driving with hands upon the wheel will give us a feel for the boat in all sorts of conditions and make us better sailors.

Weeks ago, I maneuvered *Wild Hair* into the slip; tonight it's his turn to take her out. I set up a neat little trick with the bow line—running it the length of the boat to attach on the farthest-out piling—so I can maintain control as the boat backs out. Keeping control of the bow, I will push off the splintery post to facilitate the nose turning against the current.

At precisely 01:00, the engine is warm, Dave signals he is ready behind the wheel, and I cast off lines.

We guess wrong. My trick with the bow line is ineffectual. Playing the hand we've been dealt by the water, wind, and propeller, my partner improvises a looping, ridiculous backward route. The bow refuses to swing clear of the pier, the stern glides beyond control toward the prow of a shiny new yacht. Dave increases the boat's speed to move more river water across the rudder and gain steerage.

Then he rotates the throttle forward, spins the wheel, and turns in the wrong direction, away from our course. I understand his first priority is gaining distance around our hull from fixed objects. Sometimes using my arms, sometimes using my feet, I fend *Wild Hair* off stationary pilings and stainless steel rails that jut toward our hull like knights' lances. It's an ungraceful contortionist routine. With some speed and room to maneuver, Dave avoids cracking the shells of surrounding mega-yachts as he turns the boat about face. But on the last bend, *Wild Hair* levers noisily against a proud pier corner. It is a death rattle worthy of an operatic diva. Luckily, there is no lasting damage.

Dave accelerates in the outflowing current. I shake muscles to release tensions from physical battles with fixed things and fill my lungs with the steamy night. Eucalyptus is fragrant. I'm relieved the worst is behind us now. As Dave drives, I retrieve piles of untidy dock lines, make neat loops, and secure the ropes on the transom rail for our Bahamian arrival. The inflatable life jacket chafes my neck as I move.

This night is creepy. The route is obscure, illuminated only by the occasional glare of sweeping headlights or a random porch bulb. When I drove *Wild Hair* up river last month, the corridor was as wide and clear as a church isle before the bride. I recall paying only half-attention as I eyed boats and homes at the water's edge. But now, the corridor is crowded by shapes I cannot name. Shadows abound. Dark water is invisible on a black night so it feels as if we're slipping through space and the sensation is spooky, disorienting. It's like floating through a haunted house on Halloween night.

It occurs to me the handheld spotlight will remedy sightlessness and dispel my anxiety. The beam shoots through the gloom to illuminate a dresser mirror inside a mid-century ranch home's bedroom. My finger jerks from the trigger with the recoil of a cobra. My wrists tilt down to center on the liquid lane. The beam finds our route.

"Is this helping?" I ask, turning my head Dave's way.

"Yeah, a bit," he confirms. Looking forward again, I catch the beam passing through a second story window on the opposite shore.

"Crap. This is touchy."

"Don't worry about it," Dave says. "We're nearing the first bridge. Would you ask 'em for an opening?" We're approaching a lit intersection of street and river, and I shut the ray down.

Yesterday, I telephoned the five bridges on tonight's route and asked each tender if someone would lift the bridge deck in the middle of the night. If the streets don't rise, the first twenty-four-foot-high span would snap *Wild Hair*'s mast like an asparagus stem. Everyone assured me that an exit in the dead of night would be problem-free. I hail the first tender on VHF channel nine—Florida's bridge frequency.

I anticipate a crackly response: "Roger captain. Why don't you hang back a bit? It'll take just a moment to stop traffic and open." But, there is no reply. I try a second and third time. The radio is mute. Dave eases the throttle to slow our approach.

Here's an important thing: to maintain steerage, a sailboat must move faster than the surrounding water. If it doesn't, the boat is no different than a meandering leaf that turns and swirls in eddies. It too will drift from rock to shoal. When Dave cuts *Wild Hair*'s speed to avoid slamming into the bridge, the outflowing river takes hold of our keel and sweeps us sideways into a bank of pilings. We arrive with a thud.

All right, I think. At least here we are stationary. I consider tossing a line onto a piling but the boat is wedged tight by water pressure. We are safe for the moment.

I climb through the companionway to get to the bookshelf. Coffee mugs hanging beneath the steps clink when I bump one with my foot and set them all to swinging. This time, I'm searching for the bridge phone numbers published in the boating guidebook.

"Oh, yes," the woman mutters when I phone and tell her our location. "Um, right. Yes! I'm sorry I didn't hear your call. I, uh, was asleep on the floor downstairs."

Rrrriiiiiinng—a warning bell shrills as the traffic gates drop. The street cracks in the middle, separates like twin flaps on a giant cardboard box. *Wild Hair*'s engine powers the ship from the pilings and through the gap. Switching from cell phone to VHF, I say, "Thanks for your service bridge tender. This is sailing vessel *Wild Hair* and we are clear. Have a good night."

More silence. Doubt nudges my confidence. Is our radio working, I wonder?

Repeated cell phone and radio calls with the tender uncover the source of the problem: the newly installed VHF radio has two channel nines—one for "weather," one for "calling." But knowing this and switching to the proper channel doesn't help. I must phone and wake each bridge tender en route to the sea. Only the train bridge, which stands open by default, is ready for our passing.

I worry. Ft. Lauderdale is *the* staging ground for Gulf Stream crossings but Dave and I appear to be the only sailors in the city departing on this last opportunity to get to the Bahamas before Christmas. Have we misjudged the situation? A fleet of travelers would bolster my confidence, but we are alone.

"Why do you think we're the only ones out tonight?"

Dave shrugs. "Don't know." Hand over hand, he spins the wheel through the next river bend. "But you know what they say: every captain makes their own decision. Avoid group think." He smiles. "So, captain, what's your decision? Do you think tonight's the night?"

"Yes," I smile. "By all accounts, things should be good out there."

"Then what do we care if everyone else stays home. Besides," he adds,

"I like our plan, too." His knee nudges mine, bringing a grin to my lips.

After clearing the fourth bridge, we arrive at the place where the New River meets the ICW. Here, the meaning of signposts marking deep water flips; outbound on the river, we took red channel markers on the right and green markers on the left. When we merge with the ICW, the green markers go on the right and the red on the left.

"Heather, that channel marker there . . . right or left?" my colorblind husband asks. I wasn't following along on the paper chart. I stand and look desperately for a recognizable landmark.

"Uh, I don't know!"

"Pick one!" Dave urges. "Uh, right. Take it right."

Wrong answer. *Wild Hair*'s keel catches ground and the boat rocks forward to a sluggish stop like a bus breaking sharp in traffic.

This is a disastrous turn of events. We're stuck in the epicenter of Ft. Lauderdale's boating industry. At dawn, private and commercial sailboats, fishing boats, trawlers, old-fashioned river queens, jet skis, and mega yachts will fly from their perches, like the swallows of Capistrano, to pass through this junction. The outflowing tide that drew *Wild Hair* downstream all night will continue to drain. Our boat will tip over on its keel; the mast will scissor toward the water and halt traffic like a gate blocking the street before bridge deck rises. We will be here for hours, apologizing for disrupting commerce and trade. We will miss Christmas.

But Dave has other intentions. He guns the engine in reverse, turns the rudder hard, pushes the throttle into forward and swings the rudder in the opposite direction. Slowly, the rudder walks through the settled muck of urban runoff. The keel plows sludge. City lights illuminate plumes of green-brown slurry stirred by the prop; they erupt like time-lapsed images of cumulous clouds. Dave persists; the boat claws from bondage. Suddenly, *Wild Hair* is free and moving easily again.

"Hooray, Dave! You saved Christmas!" I lean across jackets and a flashlight to kiss his cheek. Short whiskers in his beard jab my lips.

"I was not about to be stuck *there*. No way," he says. "Now I need help with these channel markers." I grab the paper chart and together we navigate the crossroads and clear the final bridge. At last, we are free from obstacles, pointed to the safety of the deep ocean.

We pause for a moment so I can lift the mainsail and capture a boost of wind power as we motor the distance. Then, Dave turns the bow toward the red and white safe channel marker positioned just outside the Port Everglades entrance. This marks where ocean waters run deep and depth is no longer a concern to ships. I busy myself putting away the spotlight, rain jackets, and other equipment before wave action spills the works. Dave navigates the center of the harbor channel.

"I am so glad to be out of the river," I say. "We should have popped for a marina near the mouth to avoid all but the last bridge. A quick exit would have been so much easier. I knew it would take a couple hours to navigate, I didn't realize it would be two hours of terror."

"Absolutely, next time" Dave confirms. "We had no business undertaking such a technical challenge in a sleep-deprived state. Not to mention, it's really dark without the moon. We could have gotten into even more trouble!"

Neither of us hears the *sécurité* radio call, a warning to navigation issued for our benefit.

"Yeah," I admit, "that was technical. I . . . ." A bright ray sears my retinas and arrests my sentence. My mouth hangs open, hands stop, shoulders turn, eyes widen. A giant ship bears down on us, directly in front of *Wild Hair*'s bow.

What Dave and I thought was the red and white safety buoy is actually an oncoming freighter. The pilot of the ship probably saw us cruising down the centerline and issued the *sécurité* on the radio. We didn't answer or change course, so the crew is buzzing us with a beam of electric daylight to get our attention. Dave spins *Wild Hair* ninety degrees to starboard, drives full speed to the corridor's edge, and narrowly avoids a collision.

"Uh, freighter entering Port Everglades," I stammer into the radio. "This is the outbound sailing vessel, *Wild Hair*. Uh, we regret getting in your way. Thank you for making your presence known. *Wild Hair* is standing by on channel one-six." There is no response.

As we enter the deep ocean, Dave spins up *Wild Hair*'s radar to spot ships at a distance in the night.

I sit hard on the cockpit cushion and shake my head.

"I can't believe all the problems we've had in so short a time: the mess with the starter solenoid and autopilot, our ridiculous departure leaving the slip, troubles with the radio, bridges refusing to rise, misreading the channel marker and running aground . . . "

"And don't forget our obliviousness to the *sécurité* radio call," Dave adds. "That's most frightening of all!"

"How could we be so unskillful? We tried to think of everything. We planned and planned this trip. Dave looks at me without an answer. "We should have done better." I stew for a moment as I snap the carabiner to my safety harness and tether my body to the boat. I thought leaving the marina, moving downriver, would be uneventful, like our arrival weeks ago. I assumed tonight would be no different.

Hours later, we arrive at the Gulf Stream and it is as powerful a presence as a mountain. I am behind the wheel, eyeing the compass, trying my best to hand-steer the course. Underwater, I imagine a long, tube-shaped aquarium filled with life soaring northward, slick as a Japanese monorail. The humid atmosphere and intense waves tell me we are in the stream, although we can't see the color change of the warmer water in the inky darkness. The wind shoves us forward at twenty knots, twice the predicted strength. The stream's horizontal platform lifts into six- and eight-foot spikes, not the promised two-foot waves.

I'm quartering peaks, riding up and over at a slight southward angle, and *Wild Hair* bucks and twists. The boat is a handful. Besides the up and down, the hull is knocked sideways in a cork-screw action similar to a carnival ride. With no visible horizon, I get seasick and toss my dinner into a saucepan standing by. Temporarily relieved, I steady my breath and relax into the tugs of gravity and motion. I continue steering.

The good news is we're in the worst of it and we're safe.

My hands arc in great sweeping motions from the ten-and-two position on the wheel. This counteracts the shove of intense waves by making cables wag the ship's rudder. Sightless, I take in the cries of wind and water and feel with my whole being a way around and through the fluid earth. In a symbiotic bond, I am fully consumed with the task of seeing with my legs and feet, listening with my gut, speaking with my hands. I am grateful for relatively benign conditions and shudder to imagine worse.

REFLECTIONS ON SKILL

I get it. The difference between navigating known waters in darkness and in light is, well, like the difference between night and day. I took two journeys on the New River—one up to the marina and one down and away—and each time the route was simultaneously different and the same, because the world around me did and didn't shift. What I failed to realize when preparing to travel the second time was how darkness would alter my sensory perceptions, that important environmental feedback would be missing. The identical route became technically difficult and far more perilous, and this changed the way events played out and the demands on me. The only real opportunity I had to avoid difficulty was to dodge the evening route altogether and start from a place nearer the sea. But here lies the crux of the matter: when things got rough it was on me to dig deep to come up with new and resilient skills.

And this is another one of those wake-up-pay-attention messages from the Great Atlantic Teacher pushing me to circle around to the question on my mind while sailing—*how am I to live in a suffering world?* An answer: prepare in advance not just for the technological, economic, or political challenge, but also for the spiritual challenge that lies ahead. OK—the idea is simple. But that lesson applied to the reality of climate change lets loose a cascade of additional uncertainties. Is climate change a complete game changer? Has it already altered the context of life? How am I to keep current of changing events as they ebb and flow when the usual sensory inputs are insufficient? How might I effectively deal with peril I can't see?

Moving through the day—walking to the grocery store, steering the boat, cooking meals, doing laundry—my half-paying-attention-mind automatically assumes life is straightforward. I couldn't count change when making a purchase if that fact wasn't at least partly true. And yet, I know in my heart that summers are longer, birds and butterflies are fewer, the forests of my childhood are gone, and more people and cars are on the roads. I can count on my fingers things that have grown taboo in climate concerned circles in recent years: driving gas-guzzling cars, eating a diet heavy in beef and dairy, clear-cutting forests, keeping homes icy in summer and toasty in winter, wasting fresh water. The list could go on. I couldn't imagine growing up that these actions could accumulate to make Earth downright *uninhabitable*. But of course, I didn't know then what I know now. No one really did. The context of life itself has completely transformed.

I, and many others, am trying to pay attention to the shift and prepare for the road ahead. But climate change is a kind of all-encompassing darkness; no one knows exactly what it will be like in advance. Scientists have a few specifics, but the climate science itself is evolving because we're having difficulty perceiving and measuring. For example: years back, scientists predicted the Gulf Stream would stop flowing. They said something called the thermohaline circulation of the North Atlantic Gyre would shut down and northern Europe would enter a new ice age. But in time, the sophistication of climate models increased. New calculations disagree over whether or not cold fresh water from melting ice sheets would shut down the Gulf Stream and are inconclusive regarding the actual effects of the stream on Europe's winters.[8]

Given all that we cannot know, how are we to function with skill?

Wild Hair's exit in the dark from the New River was a nail-biting farce. But the boat's equipment failures, wrong turns, and sleeping bridge tenders demanded something different from me. I was forced into new levels of resiliency by those struggles. I upped my focus and stayed steady, got creative, and felt appropriately humbled by my mistakes. I worked with the technologies at hand in new ways. Strategies for making progress morphed. Throughout, there was no place for anger or blame and I kept taking responsibility for our safety. Motoring down river, I kept alert and honed my skills under stress. There is no doubt in my mind that meditation and spiritual practice nourished my personal resilience years in advance and prepared me for the trial.

After attending that first meeting with Thich Nhat Hanh in Wisconsin, I kept signing up for more retreats, practicing with him whenever he visited the United States. I even went to his monastery in Bordeaux, France, and sat with him and the community there for weeks on end. Wherever I went I sat on the floor (or occasionally a chair), slept in a bunk bed, ate plant-based food, sang simple songs, helped out with chores to keep the facility running (did things like cleaned the bathrooms, swept the meditation hall, or chopped vegetables), and enjoyed long periods of personal silence. Never did the retreats feel like a vacation because I kept working with a parade of thoughts and emotions that came up. (It did however feel rustic as summer camp without the sports or high jinks.) What kept me coming back was the mind-expanding, how-come-I-never-thought-of-that before insights that either Thay would calmly lay bare or my own insights would expose.

Between retreats, Dave and I practiced meditation and studied Thay's books with a local group of the Zen Master's followers in Madison named SnowFlower Sangha. Weekly, the two of us met with this nice bunch of remarkably ordinary folks to do sitting and walking meditation. As a group, we did our best to figure out ancient teachings and apply them to daily living.

At home, I put a four-by-six-foot marker board on the floor of my office. Most days before going to work I'd spend a half-hour meditating in front of it and another half-hour sketching—in blue, red, and green marker—the core Buddhist teachings I was reading about. Again and again the board filled with notes and arrows (lots and *lots* of arrows because everything seemed so connected). I'd erase it and fill it again. Off the cushion, I moved through my days as if the context surrounding my existence had transformed. I became absorbed into the texture, sound, taste, and feel of experience. I saw ordinary happenings in a new light. Looking deeply, common things held wonder, and often I felt like a baby holding a ring of keys for the first time. My powers of concentration sharpened and intensified. I came to more fully appreciate that my words and actions had direct consequences, so I took greater responsibility for them.

Although my inexperience with the altered reality of night contributed to an evening of blunders on the New River, years of spiritual practice helped me let go of preconceived notions of what *should* be happening and glued my awareness to moment-to-moment developments. Mindfulness helped me act with skill when things went cockeyed.

Now, during times when I'm off the boat, I use my marker board to track climate disruption—its physical and spiritual causes and conditions. I read and study articles and science reports and sit with the implications. In this way, I open myself to a new reality. As much as I'd like to, I can't make the genie of global warming go back into the bottle. I have to take responsibility for contributing to it and try to give life to something new.

The invitation is for more of us to build up our muscles of spiritual resilience in advance of immediate danger by practicing mindfulness, developing concentration, making the connection between our actions and their consequences, taking responsibility of our own unskillfulness, and looking moment-by-moment for new and helpful ways of acting. Cultivating a spiritual skillset in advance is the best way I know to prepare for future uncertainties. From experience I know it can be *the* difference-maker in helping transform in an instant my usual ways when things turn unusual.

Here is a poem I wrote and recite often to enhance my concentration and cultivate spiritual resilience. Each line can be practiced with a group in a call-and-response format.

Senses skewed
Context shifted
There is both same and different in the dark
How can you be

Mindful
Concentrated
Resilient
And mindful some more

Breathing, stepping, never blinking
Every action contains a consequence
Finally, responsibility is born and
Insight becomes the boat to the far, safe shore

Mindful
Concentrated
Resilient
And mindful some more

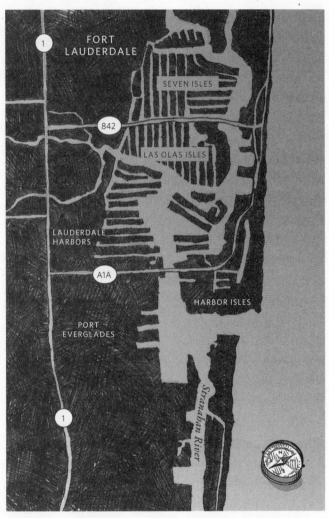

PORT EVERGLADES, FT. LAUDERDALE

4 HAPPINESS

"That's it! When in doubt, let it out," Peter coaches.

Wild Hair tosses on the watery knobs kicked up by Saturday afternoon boat traffic. I step halfway from the cockpit and look straight up at the shape of the main sail. The back of my neck pinches from the angle. The mast points toward gray, low-hanging clouds. The air is heavy and sluggish. The deck dips sideways and I grab the winch at my knee to stay steady. I'm feeling irritated over the obliviousness of some motorboat captains—those who don't know or care about the effects of their ship's wake on a sailboat.

"Let the line out until the forward edge of the mainsail flutters," says Peter. "Then pull the sheet back in until the ruffling stops. That is perfect trim." Dave executes the salesman's instruction. The stark white sail gleams in sharp contrast to the smudged sky, making my eyes squint. Dense threads, which were hammered together, change shape and crackle as the expansive plane puckers, flutters, and grows smooth. Dave is playing with the controls. *Wild Hair* trudges through the saltwater slop.

Months ago—after spending Christmas with family in the Bahamas (and terrifying ourselves silly off Tilloo Cay on New Year's Eve)—our mainsail split. The tears were small at first, so they were no big deal as we traveled the island groups of Abaco, Bimini, and most of Exuma. But anchored in the pristine waters of the Exuma Land and Sea National Park, I had to hike the mast and tape what had turned into foot-long holes near the leading edge—first vertically then horizontally. In Nassau, we had to buy more sail tape to try and stem even longer splits in the sail near the end of the boom. The fifteen-year-old cloth was done in by ultraviolet light, sea salt, and age. The fabric sagged. So when we came back to the States, we hired Peter Grimm to dress *Wild Hair* in new garb.

Now, where tired cloth once drooped, the new sail stands stiff and thick. Sail engineers shaped the triangle like an airplane wing to force the stream of air on the backside to move faster—at a greater distance in the same amount of time—than the puff of wind across the front. Breeze working against an asymmetrical plane draws the boat forward. The physics of it matters. Peter is here to explain the principles of proper sail trim so we can optimize the benefits of our investment in advanced technology and sail faster.

Wild Hair seesaws as a private fishing vessel and a dive boat brimming with tourists zip across our path.

"We're not going all that fast," I complain.

"What?" Peter's head tilts as his eyes meet mine. He reaches to point overhead. "Look at that sail. Look at it. It's perfect. It's happy."

I scan the expanse. An approaching motor yacht pulses a Heavy Metal classic.

"Heather, this is what a happy sail looks like." He steps nearer to make his point. "This is all you're going to get from the boat in today's conditions. This is what a great sail on a great boat on a great day looks like."

I squint; my upper lip curls at the corner. I am as determined as a Silicon Valley entrepreneur to memorize the form that works so I can manufacture it on my own. But absolutely nothing looks unique about the presentation. It's ordinary. I exhale, exasperated.

Suddenly, I catch on: ordinary is perfect. I just didn't know it.

I'm experiencing Nirvana—like many days in the past—but I'm not satisfied with things as they are, with slow progress. I long for something else. Out of the blue, my long-held obsession with speed takes shape in my awareness; it looks like a little green monster.

"Oh, hello gremlin," I mutter, laughing privately. I look up again. "And thank you, beautiful sail, for your innate wisdom. You've taught me something, again!"

This is one of many examples in recent months of my learning to trust this boat. As a matter of fact, it's beginning to dawn on me that the sail's greatest gift is its voice, the way it coaches me while the boat is underway. Silence (I now see) means everything is A-OK. A hushed "dut—dut—dut" fluttering along the leading edge whispers the wind is shifting forward; I must trim the fabric, pull it tighter to the air. "Bang-snap!" describes a fickle breeze coming from near the bow, indicating it's time to alter course and tack away decisively from the wind source. A molar-jarring "THWAACK-rattle-rattle" reports the wind is gone. Without the stabilizing force of a sea breeze, when all is slack and rocking, the equipment is at risk of damage. Then, I drop the sail, secure the boom, and motor. But today, the sail spoke to me without making a sound.

"Got it?" asks Peter.

"Yeah, I've got it," I say, smiling and shaking my head. "It's my problem, not the boat's."

I've always equated slow progress with doing something wrong. For some reason, I equated speed with talent, so I wanted to go fast to prove competency. Slowness undermined my confidence and—on a needling, subliminal level—my sense of worth. It never occurred to me that I couldn't move faster than conditions allowed, that boundaries to my idea of success lay outside my control. This unconscious union of speed and self-image kept me from enjoying many peaceful present moments. Craving for impossible outcomes made me feel incompetent when instead I could have relaxed into the languor of the prolonged and the gift of the easy. The tension across my shoulders loosens and I notice the two couples on the nearby trawler sharing a joke.

This is the neurosis of doubt. I remember once Thich Nhat Hanh called doubt the saddest of all human conditions because it penetrates and injures relationships, and robs us of freedom and ease. The subtlest version of the poison of hatred and aversion, doubt obstructs our view of what is right and beautiful in the world. It provokes restlessness and feeds a craving for more and different. I know from experience, however, the instant I catch myself in doubt—see the little green monster at work—I release the poison, and happiness floods in to take its place. Even if I'm fundamentally dissatisfied with the reality of the moment, the insight stirs acceptance and contentment.

I visualize my green gremlin turning into dust and blowing away.

"Look," Peter explains. "You can work this thing to death if you want—jacking the mainsail up and down by inches to add or subtract shape, moving it in and out with every wisp of air—but you're cruisers, not racers. You might get another quarter to a half-knot of speed on the boat, but it's impossible to do that all day, every day. What's your hurry? There is no contest. You'll get there."

He's right. I don't need to rush. I can slow down and experience the natural order of things. I can find comfort in the layers of wilderness operating simultaneously all around us, the nesting of ecosystems. My body is host to bacterial life forms; my boat—filled with tools, water and food from land—is host to my body; the ocean, sky, and surrounding life forms are host to my boat; planet Earth is host to the ocean as it spins around the sun; our galaxy in the expanse of space is host to Earth and this solar system. Bacteria, boat, ocean, sun, galaxy—everything is working in harmony to sustain me. I am cradled by the universe, whose elements conspire to keep me safe. Seeing this, I feel immense gratitude to the whole. I am deeply happy.

Today's sailing lesson from the Great Atlantic Teacher is deep and simple. I craved speed in the midst of perfection. I suffered from doubt and the belief that I lacked talent. Peter—a skillful messenger—revealed the problem was rooted not in external conditions but in my perceptions.

This insight extinguished my craving for more. Now I am happy. Dave drops the mainsail and I spin the boat toward port. The wheel is cool in my hands. Queen's "Bohemian Rhapsody" plays in a bar on shore and I add my voice to the song. Arriving at the pier, Peter steps off. He is rushing to meet his son.

"Thank you," I blurt. "The time you spent with us today made a real difference in how we will do things from now on." Words catch hard in my throat and I grow speechless. He has shown me perfection, brought my gremlin of neurosis into daylight, liberated me from doubt, and helped me see reality. Finally, my ego can end its struggle to prove its worth when sailing. I can relax into the pace of the natural world and feel satisfaction moving at the tempo of the moment—whatever that tempo may be. I am grateful for Peter's teaching, but his mind is unaware of the profoundness of his own message. There are no words to properly acknowledge the magnitude of his gift.

REFLECTIONS ON HAPPINESS

On several occasions I've heard Thich Nhat Hanh say happiness is humankind's natural state of being when doubt and craving cease. This implies happiness is a usual thing—as ordinary as a bathtub filling with air when soapy water drains. He also says the purpose of life is to be happy. The first time I heard him make these statements, my mind stopped. What? Wait—I thought the purpose of life was to be *productive* or *important* or to change the world in some unimaginable way. It's OK for me to be happy? Just happy? Me? What right do I have to even *be* happy when the world is so hard for so many?

Teachings on happiness are surprisingly difficult for me to accept, and I don't always feel the rightness of them in my gut. After all, some people suffer from depression due to chemical imbalance: isn't that their "natural state"? Others got the short end of life's stick and live in spirit-sucking conditions of poverty, oppression, and illness. And even though my mainsail gave me the direct experience of happiness flooding in once my craving for something different evaporated, there's got to be more to this story. I want to dig deeper because happiness seems like an awfully important thing, if it is the purpose of life and all.

First off, I think it's critical to know what Thay and other experts mean when they talk about happiness. Even scientists in the growing field of positive psychology admit definitions for the word are fuzzy and subjective. What they can say with certainty is what happiness is not. It isn't a new car, job promotion, once in a lifetime vacation, thrill ride, or the best chocolate soufflé you've ever tasted. No, happiness is not a temporary or wavering human emotion. They say happiness is something steady, a deep personal contentment some people have, a gladness for life. When I think about being happy, the sensation of feeling peaceful and completely at ease in my own skin comes to mind. That's what Thay says is the happiness of humankind's natural state.

Researchers go on to say money and happiness do correlate, but only to a point. When we are truly poor, our freedom of choice is limited, and freedom is a necessary contributor to personal satisfaction. It doesn't take much imagination to know helplessness, passivity, despair, malnutrition and the injustices that accompany deep poverty erode the human capacity for gladness. But, here's the thing: the corollary relationship between happiness and wealth falls apart at the annual income level of about $13,000 per person (on a global average).[9] In other words, it is possible to be deeply happy as soon as we have food in our belly, clothes on our back, and a safe place to dwell. Everything else Americans lump into the pursuit of happiness (the goods we shop for, stimulations we seek, destinations we visit, physical comforts we enjoy) is nonessential gravy and not the food of happiness. These findings are more than counter intuitive; they're counter culture. Yet researchers insist that once basic needs are met, money doesn't buy happiness. People grow unconsciously accustomed to their standard of wealth at lightning speed. The human pattern of wanting-acquiring-wanting-acquiring ad infinitum is so well documented that it has a name: the "hedonic cycle." This is why experts say more money (or more things money can buy) doesn't equal more joy.[10]

I think of the Buddha as the earliest pioneer breaking ground on the workings of happiness. In a breath-taking departure from the spiritual leaders of his time, those who spoke of happiness as something restricted to the afterlife, the Buddha said not only do each of us have the power to be happy here and now, but that happiness is the whole point of life. I imagine this teaching landing on the Buddha's students and then each generation of men and women in turn (yes, Buddhism was from day one a co-ed sport) saying, Wait—what? Then one day a young boy wanting to be a monk in Vietnam heard the message, and spent decades testing the claim on the hot coals of life. Eventually, he and I crossed paths and the elder transmitted the teaching to me. Mine is the first generation to be able to flip on the computer and check the claim against objective research; scientific data backs up the lesson.

So these are the facts according to the Buddha: all people suffer because suffering is just part of the human condition. But suffering has roots and causes, so it is possible to transform the roots, and thereby transform our suffering. What's more, we can live in ways that avoid future unhappiness. These, he said, are the Four Noble Truths: we suffer, suffering has a cause, suffering can end, and there is a path to avoid suffering. This teaching is the cornerstone of all Buddhist inquiry.

The disassociation between money and happiness works at the level of countries, too. National Public Radio correspondent Eric Weiner wrote after researching the happiness of nations, "the United States is not as happy as it is wealthy."[11] You can say that again. Americans suffer mightily. The Center for Disease Control names suicide the tenth leading cause of death in the United States;[12] two out of every three people in the US are clinically overweight or obese,[13] making us one of the heaviest nations in the world;[14] we have the highest rate of incarceration;[15] only five percent of the world's people live here but we house more than 20 percent of all prisoners;[16] nearly half of all marriages end in divorce;[17] more than ten million people are harmed by domestic violence every year;[18] one in five women and one in seven men have been raped;[19] the Unites States' homicides-by-guns rate is twenty-five times higher than the *combined* rate of the *twenty-two* countries most like us;[20] one in twelve adults, more than seventeen million people, abuse or are addicted to alcohol;[21] Americans—almost 20 million of us—use two-thirds of the world's illegal drugs;[22] almost 10 percent of us ages 12 and up used an illicit drug in the past month.[23]

Unfortunately, America's unhappiness is even deeper and more damaging than these statistics depict. People's suffering cannot be confined to statistics. Like tragic Shakespearean characters, we pine for particular expressions of love, recognition, safety, and symbols of success. Longing melds with doubt to produce sustained unhappiness and dissatisfaction. We bumble along, never really knowing what to do to make ourselves happy.

The average American numbs malaise and softens life's disappointment not by suicide, violence, or drugs, but by shopping. I get dizzy when I consider how we lead the world in per capita consumption. Americans are only five percent of the global population, but we produce half the solid waste on Earth. We devour one-fourth of all the oil and coal, one-third of the world's paper, and enough excess calories to feed an extra 80 million people. As a nation, our actions are the least sustainable on the planet. Yet we are the nation least concerned for our individual and collective environmental impacts.

Somewhere along the line, humans came to believe affluence and its trappings are *the source* of happiness. Confused, more and more members of our species are trying to satisfy an insatiable want in a never-ending hedonic cycle. Today, we've organized ourselves so brilliantly around the making and selling of stuff that we're quickening our own demise. In years past, we might have dismissed superfluous supply and demand as benign, or even applauded it for contributing to the economy. (In the aftermath of 9/11, Vice President Dick Cheney went so far as to say shopping was *patriotic*—a way to express national confidence and poke terrorists in the eye.) But we know enough now to make a clear connection between the fossil fuel energy spent making and shipping nonessential goods, the voracious exploitation of finite planetary resources and habitats, and the greenhouse gases trapping the sun's rays and heating the planet. Looking at climate change through the lens of the Four Noble Truths, we see humans suffer; suffering grows and harms all beings as we exploit Earth's life support systems; it's possible to stop the exploitation; and—if we do—happiness will come to ourselves and others. Understanding humankind's malaise and deep unhappiness as a root cause of global warming, we step onto the path leading to happiness. Put another way, happiness is an antidote to climate change.

The invitation in front of us is to get happy like it's our job. Earth needs humankind to be happy and satisfied in order for the hedonic cycle of destruction to stop. When we look, Nature is teaching us we do not have to wait to be happy. We can witness the beauty and comfort of the natural world, the way our lives are cradled by the whole, and we can cultivate a gladness for life in ways that do not harm others. We need to learn how to be happy with what we already have while swimming in the midst of the mess we've created. Again, Nature shows us the way—the lotus blossoms in mud, tomatoes feed upon rotting compost. Climate change itself is the opportunity for transformation; the sickness can be the medicine.

The good news is that for those of us not suffering from clinical depression or other mental illnesses, happiness is a choice, a practice, and an art. It begins when we wake up to our unconscious dissatisfaction, accept what is, reverse our behaviors perpetuating *un*happiness, and choose to act in ways that manifest joy and well-being. We can bring suffering into the light of day, get to know it better than our best friend and then act in ways that bring peace and healing to ourselves and others. When we are no longer unconscious victims of our doubt and wanting, we may discover mental and emotional freedom to take in all that is wrong and—just as importantly—all that is right. Taking nourishment from that which is healing is central to the art of happiness-making. In this way we intentionally cut off that which feeds suffering and bring joy and well-being to ourselves and the Earth.

Here is a meditation I wrote and enjoy often that helps me let go of things that bring suffering, and take in all that is nourishing and joyous in my life.

Breathing in, I see what is beautiful, good, and true in my life.
Breathing out, I smile to what is beautiful, good, and true in my life.
In, seeing beauty.
Out, smiling.
(Ten breaths)

Breathing in, I see the ways in which I suffer in my life.
Breathing out, I extend compassion to my suffering self.
In, see suffering.
Out, extend compassion.
(Ten breaths)

Breathing in, I see unskillfulness in my actions when I am suffering.
Breathing out, I see the way my suffering harms others.
In, unskillfulness when suffering.
Out, harming others.
(Ten breaths)

Breathing in, I see the root causes of my suffering.
Breathing out, I see how I can stop feeding
     the root causes of my suffering.
In, see root causes.
Out, stop feeding roots.
(Ten breaths)

Breathing in, I see simple, healing actions I can take
     to bring happiness to myself and others.
Breathing out, I aspire to bring happiness to myself and others.
In, see healing actions.
Out, aspire to bring happiness to self and others.
(Ten breaths)

Breathing in, I feel happy in my body here and now.
Breathing out, I smile to happiness.
In, feeling happy.
Out, smiling.
(Ten breaths)

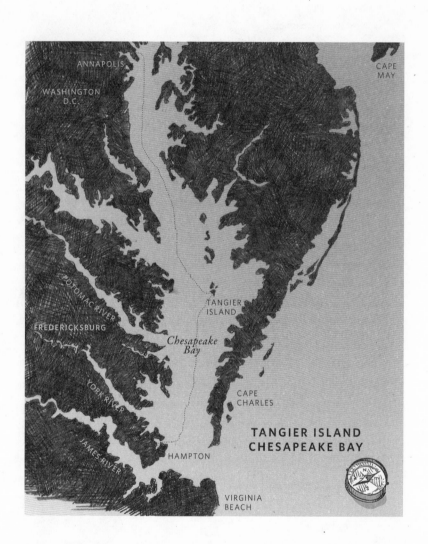

5 ETHICS

For a while, *Wild Hair* appeared to be getting younger by the day. When we were docked in Florida—the epicenter of boats and boat expertise— we installed new rigging to hold the mast tall and a new roller furler— the spinning drum near the bow that rotates the forward sail upon itself for storage. Dave—bless him—rebuilt the heads. He also put in a fuel polishing system to filter naturally growing algae out of the diesel and installed a motor mount on the transom to hold the outboard engine. I taught myself all about how radio waves move through space, bounce from the ionosphere, become distorted by sunspots (magnetic fields fluctuating on the surface of the sun), and the dos and don'ts of FCC licenses and laws governing my use of our new single-side-ban radio; workers came onboard to put up the ham radio's antenna. The real difference- maker, though, was the bright work. I stripped *Wild Hair*'s teak inside and out and lathered on ten coats of captain's varnish between the sun- shine state's afternoon cloudbursts.

Eventually, Dave and I could take crawling on our hands and knees and sanding our fingers to the bone no more. We pointed *Wild Hair*'s bow north and stitched our way outside on the ocean and inside on the ICW, out and in nearly the length of the East Coast. Every night we'd step off our ship in one of America's historic port towns: Beaufort, New Bern, Ocracoke, and Hatteras, North Carolina; Hampton, Mobjack, and Deltaville, Virginia; Christfield, and Solomons Island, Maryland. For seventeen days we were residents of Annapolis and we attended the boat show—the nation's largest in-water sailing shindig. There, I had a once-in-a-lifetime chance to meet and thank the women who inspired me to set sail: author and circumnavigator Beth Leonard; *Cruising World Magazine* editor Elaine Lembo; Kathy Parsons and Pam Wall—two of the founders of WomenandCruising.com. Later, Dave and I rented a car and drove to Washington, DC, for dinner with our son, Eland, before his midterms started at George Washington University.

Months of being on the go were hard on *Wild Hair* and a new to-do list grew. In Annapolis we fixed the electric toilet, alternator belts, brackets, and bolts, the nema cable powering our chart plotter, and our dinghy's outboard that for some inexplicable reason suffered from chronic choking and sputtering. More significantly, the mainsail had pulled from its track and now we couldn't fully raise or lower the canvas. But what really affected the quality of our days were the four port lights and four hatches that took to leaking. We froze our butts when fall temperatures dropped suddenly to thirty-eight degrees and it rained for five days. The two of us slept under a tarp to stay dry, but I felt helpless as I watched my beautifully restored woodwork pucker and stain with puddles. Worst of all, we have no way to repair the leaks until we're back in Florida in December.

But today it's dry and sunny and we're on the move. Long light beams stretch from the far horizon to illuminate the surrounding gloom of nearby clouds. Six miles south of the Virginia-Maryland border, patches of orange and yellow dot the scrub of Tangier Island. The air is crisp and cool. *Wild Hair*'s bow spins toward the buoyed entrance to the island and my imagination ignites with the imagery of *Chesapeake*, James Michener's historic epic. I feel an affinity with Tangier's first European visitor in 1608, John Smith. I'm too distracted by the history and beauty of this place to pick up signs of the modern islanders' difficulty. Besides, islanders camouflage the evidence of their pending demise behind civic order and pride. I cannot know that I've sailed to the frontline of the climate crisis.

Mailboat Harbor—the central waterfront of Tangier's quaint fishing village—is storybook perfect. One-room structures, whitewashed clean, rest atop a network of slanted piers and pilings. Fingers of reedy grass wind through the water. Electrical cables and tilting utility poles add a loopy connectivity to the infrastructure, shoelacing the whole. Between neat shacks and suspended walkways, crab pots lie in meticulous clusters: four high, five wide, and a dozen deep. Professionals readied the cubes. They are identically stuffed with folded line and wooden floats painted pink, green, and yellow. Traps are poised for business. I smile, impressed by the mindfulness of the people who created this order.

Radiant chalk-white workboats with names of women float in the gaps. The nearest boat has three VHF radio antennas, the neighboring boat has five, and none have less than two. Radars, floodlights, voluminous decks, and sturdy pilothouses for comfort and safety in foul weather are also standard. A few ships have empty stacks of bushel baskets. These are serious, expensive workboats—yet nothing is secured from vandals.

Humans are absent, but the air swirls with seagulls and willets. A cormorant drives forward submarine-like on the surface, its head swiveling on a periscope neck. Blue heron hunt for supper, their tall legs camouflaged by golden grass. A pelican dives headlong into deeper waters.

Inside the protective wall of granite boulders, we maneuver *Wild Hair*'s port side toward the pier of Park's Marina, the island's only facility open to visitors. Late in the sailing season, we are the only guests here. An elderly man arrives in time to catch the bowline, and deftly cleats the rope.

The man walks the pier in the midst of a swarm of animals. A large dog—a bale of cotton on four legs—circles and skips underfoot with six tussling cats. I've never seen a dog and cats play with such merriment. How can the man stay upright in the center of the circus? The effect is magical, like suddenly coming across St. Francis of Assisi in the middle of his day. The dog darts between the fellow's legs and worry for the old man's safety knits my brow. I hope he can swim, should he trip. I pass the dock master the stern line so he can tie us off and keep watching. The animals' darting and pawing has an everyday appearance. They are playful, but bored. The man is untroubled by the cacophony; he must have made peace with the din years ago.

Our kitten jumps into the cockpit, walks to the limit of her leash, and yowls a protest that shrivels my ears. Dinghy the Sailor Cat, a six-month-old gray American Shorthair with a white face and bubble-gum-pink nose, is the newest member of our crew. Getting her was an uphill battle. I'm a pet person, you see. Life didn't feel complete without a cat or dog. But Dave thought an animal onboard would be too much of a bother and maybe even a danger.

"Heather," he said when we were in Wisconsin last May to help Maggie move between college apartments. "If we have a pet I'm afraid you'd save it in an emergency instead of yourself—or worse—instead of me."

"Don't be ridiculous. I can keep my head."

"Well, we can't have a dog. It would tie us to shore four times a day. Or, we'd have to train it to go on a mat at the bow and I don't want to constantly clean up poop."

"I don't either. A dog doesn't make sense. But a cat—all we need is a box and the creature would take care of itself."

"I'm not a cat person."

"I know, but I am." And so we went, round and round. Until one day I put my foot down. "Dave," I said. "You are cuddly, but you're not cuddly enough!" He looked at me and blinked.

"I have absolutely no reply," he laughed. "I guess you need a cat." And just like that, the decision was made.

Cats can be notoriously aloof, so I started looking for the most affectionate feline I could find. I wanted Dave to fall in love with it. Also, we wanted a cat that wouldn't shed a lot of hair—on the boat, hair could clog pumps and drains. My online search for the animal of our dreams said the nonshedding Devon Rex breed was the ideal cat for Dave the dog lover. So, I found breeders with litters. The problem was the animals cost nearly as much as a life raft.

"Maggie," Dave said one Sunday while we were home. "Why don't you take your mom over to the humane society and see who you meet. You know your animals. You could give her some advice." He knew I'd fall in love with the first kitten I saw, so this was his scheme to save a pile of dough. Sure enough, Dinghy's pink nose did me in. Now, this cuddly and curious cat is responsible for crew morale. Even Dave has thawed to her charms. As a matter of fact, he washes her in so much affection that I recently felt compelled to institute a new boat etiquette policy: if you kiss the cat, you must kiss your spouse.

At Park's Marina, Dinghy arches and fusses to assert the boundary of her territory. But her romping counterparts ignore her claim.

On Tangier Island—our guidebook says—un-fixed felines roam in abundance. Sailors accuse locals of managing the feral population through forced adoptions; it is said islanders toss cats onto departing boats hoping the crew will offer them good homes. Suspicious, I give the marina operator another once-over. This guy is in contact with all visiting sailors, so he's probably the leading cat pusher. I tighten my frame and steel myself not to fall victim to a forced adoption.

"This is our cat, Dinghy. We already have a cat so we don't need any more onboard. Can you help us keep the other animals off our decks?" Occupied with the job of securing fenders between *Wild Hair*'s hull and the pier, the man smiles politely and nods. He continues to operate in the midst of animal mischief. My impression flips. The man is altogether unconcerned with cats, their activities or their fates. I doubt he has a population control agenda.

With the boat settled, Dave and I step onto the pier. Proprietor Milton Parks introduces himself. "I am a Waterman," he says. "I made my living like the other men here by harvesting Chesapeake crabs."

"With boats and traps like these?" I ask.

"Yes, from four in the morning until five at night, six days a week. Plus, forty years ago, my sons and I spent nights and weekends building that granite seawall by hand. That made it possible for us to construct this marina." My eyes take in the wall's football-field length. I feel the weight of a single rock in my arms and back and am astonished by his courage to even begin the job.

"That must have been incredibly difficult!"

"Yes, moving rocks and hauling crab pots is work." Mr. Parks smiles without complaint or embellishment. He continues like an educator. "I am a military veteran and a direct descendant of the island's first settlers, too."

"Then you are the best ambassador for the island—the man who can answer all our questions."

"Yes, that's true."

"Do you still do any crabbing?"

"No. I'm seventy-eight years old and retired to the marina." The gentleman's face is weathered under his baseball cap but his energetic frame is fit. He's rugged and youthful for his years.

With animals bouncing under his feet, Parks shows us the showers and the road into town. Nodding his head toward a tidy ranch house on shore, he says, "That's my house there, the only brick home on the island. My wife's inside, but don't bother her. She's busy sleeping with other men: Ben Gay and Absorb N. Junior." I'm caught off guard, and then laugh at his unexpected silliness. A sign on the shower shack catches my attention.

"Mr. Parks, you charge only $25 a night for boats smaller than thirty feet and $30 for larger ones? That sign says an electrical hook up is $5 for 30 or 50-amp service?"

"Yeah, that's right. That's what you pay."

"But," I stammer, remembering the $180 overnight charge we paid months earlier in the Bahamas, "you can charge more."

"Doesn't matter. That's the price." Two cats spar on either side of my right foot.

"All right," Mr. Parks rubs his hands for warmth. "You're going to want to eat a crab cake dinner, and you're going to want to eat that dinner at Lorraine's Restaurant. That's the only year-round establishment. Now, it's Sunday and Lorraine keeps church hours. So you do what you have to do and I'll collect you in ten minutes to take you to Lorraine's for dinner."

Dave catches my eye. "Uh, well, it's not even 3:30," I say, "and we had a late lunch. Perhaps . . . "

"No," he says. "You want to eat at Lorraine's and she's closing at five for church. I'll take you there so you can get some local cooking. There's nothing better. Then I'll collect you again and give you an island tour before dark. I don't keep church hours, you know." His eyes flash with the gleam of a rascal. I give Dave a second look. We'd be fools not to follow Mr. Parks's lead. The plan is set.

We carry Dinghy below decks, grab jackets and cash, and pocket the camera, before meeting Mr. Parks and a pair of cats at the golf-cart. It has a bench seat for two and we are three. Animals scramble as I slide to the center of the seat. Dave wedges in on my right, his hip hanging from the edge. Mr. Parks squeezes to my left behind the steering wheel and his hand fumbles around my knees.

"A dirty old man invented reverse on a golf cart," he explains with twinkling eyes. Again, the man's cheerfulness surprises me.

The cart loops in a reverse crescent. He fumbles near my knees again and drives forward toward the village. The road, wide enough for two golf carts, is a bed of crushed oyster shells. It carries us east along the shore toward something called "Pooges and the Community Dock." A turn south leads to the central spine of the island's Main Ridge. After a few moments, the cart stops at a crossroads. We're in the heart of the Meat Soup Neighborhood. Elevated white clapboard buildings with aluminum awnings sit close to the road. Lots are distinguished by picket and chain-link fences. A bicycle leans against a gate. Four cats lounge on a stoop. Small satellite dishes sit in yards like buttons on sticks. Clotheslines stand at the ready. A bench invites conversation. The scene is close, pedestrian-scale. My sense of personal space must be calibrated to the boundlessness of open space at the water's surface because my skin itches with claustrophobia from the sudden closeness of the scene.

"Lorraine's is just there," Mr. Parks indicates an even narrower lane with a nod. "I'll call on you after a bit for the island tour. Enjoy your supper."

Walking the lane, a painted sign tacked above a line of propane tanks catches my eye.

Fresh Seafood Daily
Specialties: Hot Crab Dip, Creamy Crab Bisque,
Signature Crabby Fries, Homemade Desserts
Prepared and Served by Tangier Women Using Favorite Island Recipes!
LET YOUR LIGHT SHINE MT 5:16

I am puzzled by the women who wrote this sign. We're different. I wouldn't think to advertise womanhood as the source of something especially unique in the marketplace. Also, I'd feel shy about broadcasting my spiritual beliefs out of context, assuming too much could be misunderstood. These women are bold: about crabs and Christ, they have no doubt.

I feel as if we're breaking and entering when Dave opens Lorraine's screen and I push through the front door. But as we step inside what would be the living room of the house, there are a handful of upright booths, a food counter, shelves, and a counter with a cash register and Pepsi sign. The layout is practical. The room is clean and well lit. I relax into a booth near a window.

Two women dressed casually in plaid over-shirts and jeans, with dark hair pulled simply away from make-up free faces, chat comfortably from either side of the counter. They speak with a thick local accent. Words combine hard cockney vowels and stretch them long in a Virginia drawl— it's what linguists call an English Restoration-Era dialect of American English. The dialect is Cornish and Southern, and it's unique to the island's five-hundred residents. Some claim—because of the island's isolation—residents speak the language of the seventeenth-century settlers from southwest England. Others suspect the insular community evolved its own dialect. The women's vowels are loud; they weight their Rs, and swallow their consonants. It's captivating.

I listen for particular linguistic characteristics: an ordinary word made short or the use of completely new words (like *wudget* meaning a plump wad of cash, or *coferdbent* meaning twisted). I hope mostly to detect their extreme form of sarcasm, what the locals call "talking backward." Someone might say, "You are smart," but I have to know the inflection in their voice to tell if they think I am really clever or painfully stupid. Despite my best efforts, my ears swim in confusion and I can't make out even the most basic roots of English.

Scanning the menu, I remember alcohol is not sold on the island. These are religious people, evangelical Christians. A splash and sizzle comes from the kitchen and the room fills with the smell of delicately fried seafood.

"What can I get for you?" Lorraine Marshall stands before us. She talks with the mid-American accent of a major network anchorwoman.

"Oh, I can understand you!"

"Yes," she laughs. Her look is both self-conscious and proud. "We can talk like everyone else, but we do get going between us sometimes."

The soft-shelled crabs and crab cakes that arrive in moments are golden and crispy. The edge of my fork separates a section of meat and hot, naturally salty juice mixed with a little oil oozes. The crab is firm but tender and remarkably sweet. Lorraine makes the fruit of the sea "Tangier-style" by washing away the creature's innards before cooking. This smooths out the flavor. Across the table, Dave chews with rapture, eyes closed.

I want to put a cot in the corner and move into the restaurant—never leave—but Lorraine wants to get to church. So, too soon we are back in the golf cart for our tour. Mr. Parks fiddles, barely touching my knees. He giggles anew. He is as playful as his cats. The golf cart accelerates on Main Street. We drive past one and two-story New England-style buildings dressed with brick chimneys and working shutters. Homes advertise crafts and gifts inside, signs of a modest summer tourist industry. Oil tanks and shrubs flank homes, but I see only one tree—a lone survivor of hurricanes. The cart stops in front of the church; its grand steeple connects heaven to earth. Together with the Tangier water tower, it gives the island a distinct profile.

"This is the new Swain Memorial Church. It's built where the old church stood. We've been Methodists on the island since on about the end of the Revolutionary War. Read there," Mr. Parks directs. "It's a list of church members in 1825. Do you see my name?" My eye scans the list: *Crockett, Dise, Paul, Pruitt, Thomas, and Parks.* "That's proof. I'm a direct descendant of the earliest settlers."

"I see it," says Dave.

"Before European settlers, the Pocomoke Indians lived here. The beaches are loaded with arrow heads and spear points. They wash up all the time."

"And here is the memorial," he says, redirecting our attention across the road, "to the men who served our country's military. I'm here, too. A lot of good people gave everything to defend this nation, same as on the mainland." Emotion squeezes off the flow of words; there are stories not included in today's tour. With a breath, Mr. Parks looks me in the eye. "We're Americans here." His patriotism stirs me.

"Yes, I understand." We linger respectfully in silence. Five cats lounge at the foot of the memorial. One stands and stretches. Mr. Parks moves to the cart.

Next, the buggy stops near a construction site littered with debris. "This will be the David Nichols Health Center," says Mr. Parks. "Our current clinic is a foul place. The roof and walls leak. There's no hot water. The medical equipment is troubled, held together with duct tape. But the people of Tangier love Dr. Nichols."

Canadian by birth, Dr. Nichols visited Tangier in the 1970s and found the place a living Norman Rockwell painting. Each week, for the past thirty years, Nichols arrives in his private plane or helicopter to deliver health care. He makes house calls. He saves lives. The doctor's only fault is spending too much time with patients. Once, when darkness set before he finished, locals circled the runway in golf carts, headlights blazing, to light the air strip.

"He and his friends raised two million dollars for this clinic," Mr. Parks continues. "Most of it pays for the medical building; the remainder is for care for years to come. My daughter is a doctor now, too, on the mainland." I try to imagine how far a journey it is for a local girl to live in the city, graduate medical school, and launch a practice.

"Now, Dr. Nichols is a patient himself. Years back he came down with melanoma of the eye." Cancer. Mr. Parks continues: "We are so fond of Dr. Nichols, we gave him an honorary island name: David Nichols Parks."

What we don't know now, standing at the construction site, is that, weeks before the clinic's grand opening, experts will find an inoperable metastasis in Dr. Nichols' liver. He will die five months after the news. His family will bury his ashes beside the new clinic.

Driving on, buildings give way to the vast, golden-green marshland of the island's interior. The vegetated carpet is mounded and coarse. The island is small—a little over a square mile. It's only three feet above sea level at its highest point. At the marsh, I feel the indecisiveness of water and land; it's as if the landscape can't make up its mind to be dry or wet. The interplay is sensual.

The cart turns east and rumbles over a narrow boardwalk bridge spanning Ruben's River. Canton Ridge—the low rise ahead—is the site of another cluster of homes. This is the island's first settlement.

"You see that house there?" Mr. Parks asks. A home with ample square footage fronts the marsh. "That fellow visited the island. He came by boat, like you, and he liked the place, decided to make a home here with us." My heartbeat quickens; the thought excites me.

"He is a smart guy," I say.

"Some think we islanders are too close and set in our ways, but everyone is welcome here on the island." Mr. Parks punctuates the statement with a firm nod. I imagine living in an older, smaller home in the heart of the Main Ridge, not new construction.

"We would thrive in this place, wouldn't we?" I ask Dave. He smiles in agreement as he takes in the land, water, and sky.

Losing daylight, we pass on a visit to West Ridge, the location of the airstrip and the Hog's Ridge and Sheep's Hill Neighborhoods.

"The island is void of traffic lights, ATMs, and violent crime," our tour guide says. "There's one pay phone, a hardware store, and a grocery. We may get Internet next year."

I quiz Mr. Parks on the life of a Waterman. The one-room shacks on the network of piers near the marina are Watermen's offices. The structures were built by the fishermen's grandfathers and great-grandfathers. The buildings and the necessary state fishing permits are passed to the next generation like trust funds and family jewels—through wills. The building blocks of a Waterman's business are people's most treasured assets.

There are two types of Blue Crab, Mr. Parks says: those that shed shells and those that don't. About half the Watermen catch crabs that molt, making Tangier the soft-crab capital of the world. Inside many of the shacks, aerated tanks keep crabs alive until they kick off their shells (something they do twenty times in their life). The waterman refrigerates the crab to keep new "skin" from hardening. Soft-shelled crabs require constant monitoring and a quick trip to market. This is why they fetch ten times the price of their nonmolting kin.

Watermen catching nonmolting crabs make profits from a higher volume. Each day, they clear and reset traps, pack crabs into bushel baskets, and deliver them to a wholesaler for cash. The most ambitious fishermen manage up to four hundred crab pots a day. Hauling pots through forty to seventy feet of water is wet, cold, dangerous, and exhausting work. This makes Watermen shacks the place for midday naps. But today, shacks stand as empty as a ghost town in the Great Plains. The state of Virginia stopped issuing Watermen permits. Frankly, I'm mixed up and trying to make sense out of what I've read.

Ten years ago, a fierce battle raged between Watermen and environmental advocates. Signs protesting the Chesapeake Bay Foundation—a clean water and sustainable fisheries not-for-profit—filled Mailboat Harbor. Islanders and conservationists were locked in a bitter conflict over fishing policies. Arsonists, apparently from the mainland, set fishing shacks afire. For the better part of the twentieth century, pollution from area cities and farms and the over-harvesting of fish shrank marine populations so dramatically that Watermen came to depend on one species: the Blue Crab. Then crab populations wavered. To protect the fishery, the Foundation pushed for a ban on crab harvesting during winter months. The people of Tangier feared unemployment and poverty and fought back. Watermen had heaps of practical and historical know-how and they didn't buy environmentalists' calculations. On the flip side, advocates armed themselves with ecological research that they felt outweighed concerns for the Watermen's financial reality.

Susan Drake Emmerich, a PhD candidate from the University of Wisconsin, came to Tangier in 1997 to test a faith-based approach to resource management.[24] She lived on the island, shared a Christian faith, listened, and developed friendships. She asked the mayor, pastors, lay church leaders, and the women of Tangier to endorse and help her with some research. What I most appreciate is the way she offered a new understanding of Christian scripture—one containing responsibility for the natural world.

Eventually, Emmerich held a mirror up to the community that revealed the ways in which residents' actions fell short of their faith.

Through guided self-reflection in sermons, hymns, religious pictures, Bible studies, community meetings, and one-on-one discussions, Emmerich connected islanders' love for God with a responsibility to sustainably steward creation. She united loving your neighbor with respect for civil laws. She also helped residents understand how their message to outsiders wasn't helpful to their cause.

Some islanders resisted Emmerich's message. They feared change and defended the traditional ways. This split the community. Angry and hurt, neighbors, friends, and family members stopped looking each other in the eyes. They hurled insults at one another. Misinformation inflamed strong emotions, and Emmerich received death threats.

In the midst of it all, fifty-eight men knelt on the altar in 1998 to publically and before God declare their intention to care for creation. Women of the village made a covenant with God to alter patterns of household consumption and teach children the inner workings of God's creation. Tangier residents, with their hearts in their hands, apologized to environmentalists at the Chesapeake Bay Foundation. Softened by tenderness, the organization's staff apologized in return and realized they needed to respect people's religious beliefs and cultural heritage. Watermen also met with farmers and they too entered into a stewardship covenant with God to avoid harming their neighbors downstream.

~~~

The only sound when we wake the next morning is the cry of seagulls and the murmuring of waves on *Wild Hair's* hull. The Watermen must have left hours ago; they're out there somewhere, tending traps. Dave and I get ourselves organized and walk the narrow streets to Daley and Son Grocery. I step over two cats to enter. The place smells of bleach. The brightly lit room with high ceilings doesn't have much on the shelves. We buy bread, eggs, coffee, and cat food.

"Does the island freeze-in in wintertime?" I ask the weathered fellow behind the register.

"Yup, every year." He methodically rings our goods without looking at me. "Heather, come here," says Dave as he points to a snapshot pinned near the door. A young woman, arms wrapped around grocery bags, stands outside the building. She is in water above her knees.

"What's the story of this photo?"

The grocer shrugs as he slides eggs in a sack. Styrofoam squeaks. "When the tide is high and the moon is just so, the island floods."

Later, we visit the Tangier History Museum but find the door locked. A sign on the wall reads:

WALK AROUND BACK

RING THE DOORBELL OF THE HOUSE BEHIND

A bicycle blocks our path. Men's unmentionables dry on a line. Rock music plays through screens. We're not expected and—feeling that I'm violating someone's personal space—I turn to leave, but Dave steps up and presses the doorbell.

"Come in, come in, come in!" a man's voice calls.

We enter a cave of a living room. Through dim light I see clothes, papers, boxes, books, and shoes strewn across the floor and furniture. A young man leaps forward from a black corner and into the only beam of light penetrating the window. His hand is outstretched.

"Welcome to you! I'm Ken. I'm the Artist in Residence on the island. You're here visiting." He is a lanky fellow, supremely attentive, with piercing eyes. "I've been on the island since last year and my job is to make that old house an interpretive center."

"You're having fun?" I ask.

"It's perfect for me. I have art *and* history degrees so I'm using all my skills. This is my first job." The lad beams with enthusiasm. "They really need me, too. Hang on. I'll get the keys so you can check out what we've been up to."

Ken stops en route to the museum, points toward the grass.

"That path goes to the Tangier Water Trails and Kayak Rentals. The kayaks are free to anyone who wants to paddle the rivers and wetlands of the island. There's a ton of wildlife."

"How cool is that?" Dave asks.

"I got that going. And have you seen the signs everywhere? Yellow and blue?" I recall dozens of all-weather interpretive signs dotting streets and buildings. I nod.

"We made those to get word out about how amazing this place is."

"The signs are gorgeous," I say. "Nicely done."

The museum is a modest clapboard house with small rooms. Clusters of personal effects heaped upon display cases, posters, photos, and tables give the feel of an estate sale. Walls are plastered with writing. My nose picks up mold.

"People give us things they inherited from parents and grandparents. Everyone wants to contribute." Ken disappears behind a door, and lights, along with a recording of a man's voice describing crabbing on the Bay, switch on.

"You've captured the locals' stories—"

"We got funding to record and preserve tales of island life before it's lost."

I listen and wander. Another man's voice follows; his tale of a storm makes my neck- hairs stand on end. Then an elderly woman talks of family and the treasure of lifelong friends. Island life takes form around me like a cocoon of shared purpose, memory, and meaning.

I stop at a worn geological map pinned next to a modern satellite image. In the photo from space, Tangier looks airbrushed, feathered, nearly translucent. When I compare it to the map, I see a major discrepancy: the island used to be bigger. Tangier had six habitable ridges and now there are three. Visceral panic grips my throat.

The Army Corp of Engineers, according to a nearby placard, calculates nine acres of Tangier Island disappear into the Chesapeake Bay each year. The island melts like butter in a pan. It's one-third the size it was during the Revolutionary War. Uppards, a settlement north of Parks Marina, was abandoned in 1928. Tangier's population is less than half what it was fifty years ago.

Perhaps walls and bridges are needed, like those used in the Netherlands. But another clipping reveals the small population can't afford engineering studies and protective seawalls. On Tangier, the median household income is $26,607; more than a quarter of the residents live below the poverty line. With so few residents and a shrinking tax base, leaders are challenged just to keep the sewage-treatment plant in working order. Tangier residents don't have the resources to counteract the effects of Mother Nature.

A sepia photograph hangs on the wall. It depicts a train car brimming with oysters. Once, oysters were abundant around the island, and Watermen supplied them en masse to New Yorkers. Today, oysters are scarce. A magazine clipping near the photo describes current efforts to reestablish the oyster population—the setting aside of oyster reefs.

Turning a corner, I read how the State of Virginia—after years of debate—outlawed crabbing during the five months of winter and still, the Blue Point population shrinks. In response, Watermen started fishing for striped bass with gill nets. Now there's talk of banning gill nets. In the recording, a Waterman says: "The economics just don't work anymore. I'm too young to stop working and too old to learn a new job. Am I to take my family to the mainland to work in a Wal-Mart? What am I to do? I just don't know."

"Last spring," a woman's voice offers, "not one graduating high school student wanted to be a Waterman. Not one. That's never happened before in the four hundred years since the island settled."

Above the door, a rendering of Chesapeake Bay tells the story of the water body's creation. Thirty-five million years ago, a meteor struck the Sassafras River. The collision opened the river's flow to the ocean and created the bay. Tangier and the surrounding area is still experiencing *subsidence* from the collision; the region continues its slow-motion slide into the original crater. Tangier is incredibly susceptible to flooding—as helpless as some low-lying islands in the South Pacific: Tegua (part of the Torres Strait Islands) and the Solomons (east of Papua New Guinea). The idea that this community and everything I see and admire here may soon be lost sweeps my body in a wave of defeat. I take a breath; stale air moves through my nostrils.

All of this is compounded by sea level rise due to climate change. Another poster says the ocean swelled eight inches in the past hundred years due to thermal expansion and melting ice on land. The US Army Corps calculates sea level will rise another five feet here by 2100. Tangier, a whisper of land in the Bay, doesn't stand a chance.

I look out the window at the street scene. A plastic play gym sits in an empty yard. A golf cart waits patiently for its driver. The setting is no longer quaint but eerie; I feel the whole of it already abandoned.

A mass relocation of island residents to the mainland could make Tangier our country's first climate refugees. But retreat isn't on the Watermen's plate yet. These are people of faith. God provides, they think. It is inconceivable to many here that the Lord would allow such loss. Voicing concern about the future or talking about where or when to resettle indicates a lack of faith.

Ken locks the museum door and we part ways. Daylight sparkles. The wind chaps my cheeks. Things look as they did the hour before, but nothing is the same. Tangier Island is ephemeral to me now. Neat homes, interpretive signs, Mr. Parks, Lorraine, the grocer—these are living ghosts who will find their place in history alongside John Smith. The museum warped time, bent my sensibilities. What feels real is not the clean picket fence in front of me but the same boards half submerged, abandoned, rotted, and tilting. Dave and I walk in silence.

I peer down a side street to see thirty people dressed in black outside a church. No one in the group is talking. Clearly, they assembled for a funeral. They are solemn, waiting for something to happen. One man in his late twenties lifts his head and returns my gaze. He adjusts his collar and conveys disdain for his tie. With this small gesture, I am included in the burial rites, the intimacy of the community. I feel cut by the knife-edge of love and loss. Helpless to defend myself, I keep walking.

Aboard *Wild Hair*, Dinghy is eager to jump into the cockpit. Dave clips her to the leash then lets her out. I make lunch and try to process what I've learned.

"There's no need for us to rush off, there's lots to see and do here," I say. "What do you think about staying an extra night?"

"It's getting late in the season, but we could. I agree."

"I have a crazy hankering to live here on the boat and work for the people of Tangier. We could sail from here up the Potomac and I could lobby for them in Washington, DC."

"That sounds like a slow commute."

"I know, but it would be fun and these people need support. Their future is desperate and they don't have any clout."

"I'm not sure what these folks need, other than a U-Haul."

"Yeah . . . I know." My lips pout involuntarily. Commotion erupts outside.

A large fiberglass trawler, which is captained by Mr. Parks' son Doug, ties next to us on the T-dock. The senior Mr. Parks, in a swirl of cats, manages the lines. The dog is asleep near shore.

Doug Parks is sturdily built and technically proficient. He shuts down the engine and organizes equipment, bantering with his father as he moves bushel baskets. Immediately a workboat named *Deborah* pulls alongside the trawler's outer rails. The twang of Tangier's dialect fills the air as a pair of Watermen heave packed bushels into Doug's arms. Doug shoves them toward the forward bulkhead on his deck. There is a confidential exchange of money before the locals disappear with fresh baskets and lids.

"Are crabs in those baskets?" Dave asks.

"Yep. Wanna see?"

Doug lifts a container from the pile and unbends the wire holding the lid in place. Inside, dozens of Blue Point Crabs display the full spectrum of color. Sea-green half-moon shells are five inches wide. Their forward edge is spiked. Between protruding eyes, a pair of incisors lie flat, indicating the mouth's location. Segmented arms curl forward from the shell to make the creature look like a tiny, muscle-bound sumo wrestler. But instead of hands, there are cream-colored claws. The outer-most points on the pinchers look dipped in rusty red nail polish. Arm pits glow robin's egg blue. A few crabs in the bushel are upside down, displaying the shell's white underbelly. I remember the crab's evolved defense mechanism: the crab resembles the Bay's vegetated floor to predators approaching from above, and the underbelly looks like the sky to a predator hunting from below. The underside is jointed to allow movement of the claws and eight back fins. There is also a handy, pop-top-like flap to pry the animal open and access the sweet meat within.

"Doug lives on the mainland," Mr. Parks says. "He works for a middle-man in the crabbing industry."

"Yup. Six days a week, I drive the trawler here and buy the day's catch. These crabs will be in Toronto by tomorrow morning."

Doug lifts one crab and others follow, all pinched together. A contented smile spreads under Doug's US Navy cap.

"Show her how to hypnotize a crab."

Doug pries a crab from the crowd, palms it backside down, and draws slow circles on its belly. Legs grow still and relax open; claws droop. My mouth hangs loose, lulled by the tenderness of so small a gesture.

"This here's a male." I'm surprised Doug's not whispering. "You can tell by the length of this pull tab. Ya see? The female's tabs are more like half-circles." Another pair of Watermen pull alongside and Doug efficiently repacks the sleeping crab. They exchange bushel baskets—empty for full. The locals drive off.

"So that fellow gave you only three baskets of crabs. Is that typical?" I ask.

"Oh no. Crabb'n is way down. That fellow put in ten hours of work, paid for a day's fuel and a man and earned fifteen-dollars per bushel. How can you feed a family on that?"

I'm shocked. The fish market in Wisconsin will sell me a bushel of crabs for more than two hundred dollars. I assumed the Watermen profited. But it seems crabbing is ecologically unsustainable when the harvest is large, and financially unsustainable when the harvest is small. Once more, the sensation of speaking with ghosts from another time sweeps over me. Doug and his boat shimmer with impermanence.

"Can we buy two dozen crabs?"

"Sure. That'll be eighteen dollars."

"Oh wow—what a deal!"

Dave stows the crabs in *Wild Hair*'s storage box next to extra cans of soda. "Honey, look at this." The box is alive. "How are you going to cook all these crabs with our small pots and stove?"

"I don't know, in shifts?"

I look up and three recreational sailboats motor past the sea wall and turn toward the marina. Mr. Parks starts to scramble to accommodate the arrivals. Dave looks at me. "Should we go?" *Wild Hair*'s engine has been running to charge the refrigeration system, so it's warmed up.

Dave's question lands heavy. I'm smitten with Tangier. But climate change, needed conservation policies, and meager harvests are extinguishing this culture's flame. There isn't enough money to shore up a future here. I was welcomed as a member of the community. I've attended Tangier's funeral and walked among ghosts from the past and the future. A kernel of insight sprouts: I will be of best service to these people if I leave but share their story.

"Mr. Parks," I shout above the activity. "We will go to make things easier for you and your new guests."

"You don't have to rush off—"

"No, that's OK. Thank you for being a perfect host. We've seen what we needed to see and are grateful to you for your hospitality."

Mr. Parks raises an affable hand and looks down to untie our line. Cats paw the loose rope near the cleat.

"Be well." I step behind the helm, feeling determined to do something and satisfied over the friendships that formed. Dave and Mr. Parks finish the job of casting off.

I spin *Wild Hair* about and narrowly dodge a Waterman delivering the day's catch into Doug's hands. Mr. Parks raises his gaze and signals the next visitor to take *Wild Hair*'s place.

REFLECTIONS ON ETHICS

I don't know why I suspected the men and women of Tangier might be different from people living on the mainland. Maybe it was the stories of forced feline adoptions and environmental clashes that set off flares in my head. But now I know the residents of Tangier to be moral, decent, and fair. Like me, they want to rise above past harm they inadvertently caused to restore communion with others and nature. We share the same struggle to know what to do when everything is off-kilter. Lessons on how to live handed down by our parents and grandparents don't make sense. We're confused. Facing an uncertain future, I too am tempted to double-down on my modus operandi, do things business-as-usual to take care of my own—to hell with the rest. And yet, these people demonstrated there's another option—one that involves the personal work of digging deep and finding courage to open our hearts, live our ethics, and shift our behavior.

Me *change*? That's radical and a bit terrifying, actually. And yet on Tangier, relationships on and off the island and with the blue crab population shifted for the better because a small subset of islanders—not everyone—found courage enough to live their deepest values, and their actions transformed the hearts and minds of the majority on and off the island.

Even so, the applicability of Emmerich's methods to resolving conflicts within the general public is questionable. Cross-sections of America don't subscribe to the same religion and some shun religion altogether. (There's also the risk of unskillful leadership—well-intentioned do-gooders mucking with sensitive cultural issues they little understand.) Is it possible to articulate a global spiritual ethic capable of uniting people around the world in caring for Earth, each other, and our fellow earthlings? Maybe. But I'm not all that concerned about everyone agreeing on a single ethic and choose to focus instead on how people might consistently apply to their lives the ethical teaching they already know and are comfortable with. I'd be ecstatic if people practiced the codes they knew and trusted because the ethical roots of all spiritual traditions are remarkably similar.

I'll use the ethical guidelines created by Thich Nhat Hanh as an example. In his Five Mindfulness Trainings,[25] Thay says we are to protect life, practice generosity, abstain from sexual misconduct, speak and listen with care, and consume in ways that do not cause harm. This sounds familiar, yes? Nothing about these guidelines is particularly controversial, but in my experience applying them to daily living is tough. For example, every time I do something as simple as boiling water, I kill germs and bacteria. It would seem I can't protect life 100 percent of the time. Sometimes, because human perceptions are inherently flawed, I say things I don't realize are false. And I can't always anticipate or avoid unintended consequences arising from the food, medicine, travel, and entertainment I consume. This makes the application of ethical guidelines an imperfect art. Yet, there is value in trying, learning, adjusting course, and trying again.

Society would catapult toward new territories of ecological and social justice if a significant subset of us applied our personal ethics deeply to daily living: if politicians didn't try and kill each other or exploit opportunities, but instead focused on providing policies, education, and material resources to those in need and helping people do for themselves; if industries minimized killing while seeking profit and made deep commitments to sustainable production methods; and if every person protected life by eating only half as much meat, buying locally produced goods, using less energy and water, and minimizing fossil fuel-based transportation.

I've gotten better at applying ethical principles to the fine grain of daily living these past few years through deep looking, and trial and error. I've learned to stop and ask myself before I take action, "Does my plan harm myself or others now or in the future?" In the midst of a conversation I often do a reality check and ask, "Are my words helping or hurting the situation?" Crawling into bed at the end of the day, I reflect, "Was I skillful today in protecting Earth, her people, and her creatures?" More and more, I'm turning to plant-based recipes and figuring out how to prepare satisfying vegan meals. Bit by bit mindfulness brings my ethics to life and helps me avoid unskillful actions. I practice in this way because I know deep down the truth of what Thay says: actions that create suffering—now or in the future—are wrong, and actions that bring peace and well-being to both me and others are right.

Living ethically is a subtle and deep practice that I find requires vigilance and monitoring. I make mistakes when I think I know what others will say, stop listening, deny people's reality, or lean into the self-serving interest of my tribe. I'm not using careful speech when I exaggerate my message or stir anger or fear in someone else. I'm not immune to the toxicity of my own strong emotions, and sometimes I blame others and compound tensions by spewing harsh words. Imperfect as I am, practice has made me a whole lot better at buttoning my lip before I say or do anything destructive, and I've personally experienced how the effort has made the people around me a whole lot happier, too. This is what the nuts and bolts of living ethically feels like to me.

But here's the rub: ethical boundaries grow increasingly inconvenient as conflict and what's at stake escalate. A noble end can seductively lure any one of us to employ harsh means. But unskillful actions fan flames of anger, fear, and sorrow. Conversely, ethical actions cultivate confidence and peace between people. The codes of behavior handed down to us through generations help us live in ways that manifest justice and healing here and now. Our hopes for the future flower in the present.

Humankind's common moral core is neither a superficial, sentimental nicety nor a deep intimacy we share only with close family and friends. Ethics are public boundaries with the power to save us from hurting ourselves and others, and every spiritual tradition promotes them for this reason.

Happily, in 2015, the number of spawning age female blue crabs in Chesapeake Bay increased. But unfortunately, by my math, sea level rose by a half inch and the island's land mass subsided by an equal amount, for a combined total of one additional inch of flooding in just the few years since my visit to Tangier. Additionally, wind and sea action chewed up and swallowed somewhere between forty-nine and sixty-three acres of land. There are ecological processes at work that even humankind's highest ethical standards cannot alter. A moral code doesn't guarantee we will get what we want; it won't save Tangier real estate from disappearing into the bay. But ethics can protect and honor the people and culture of Tangier and help ease the transition as islanders make a new beginning.

Recently, Dave forwarded me an article from the *New York Times* with the headline, "Resettling the First American 'Climate Refugees.'"[26] As I took it in, similarities between the featured Isle de Jean Charles in Louisiana and Tangier Island in Virginia leapt off the screen. They are two, miniscule, waterlogged mounds considered by experts to be among the places most vulnerable to rising sea level; they possess unique histories with insular cultures whose members are hurling toward rapid change; a lack of consensus frays the communities' social fabric as residents wrestle individually with the difficult decision to stay or go; both island communities have a long list of reasons to disdain government; and yet their homelands cannot be set right and continue into the future by any of the known tools in engineers' kit bags. They need help.

Reading further, it dawns on me the people of Isle de Jean Charles are a lucky bunch. The Department of Housing and Urban Development just granted them $48 million in the first-ever federal grant to relocate an entire population to higher, dryer ground. Lock, stock, and barrel.

The goal is to transplant sixty people to a new homeland, en masse, within six years, for the whopping sum of $800,000 per person. That's like winning the climate disruption lottery, big time.

Even so, the job won't be easy. The two Native American tribes living on Isle de Jean Charles haven't always seen eye to eye and their leaders have to find a way to agree upon where to go and who can join them. The twin sovereign nations are afraid to lose their claim to the land of their ancestors. They worry about being accepted in their new locale, and finding work. Three failed attempts to cut and paste the community into a new environment in the past dozen-or-so years whittle at their confidence. Those efforts collapsed under the weight of logistics and politics.

I admire the fact they are not waiting until conditions are desperate, but planning ahead, making choices, proactively sculpting a preferred future. These people are leading the way to figure out how to keep a community together while extracting itself from flooding, crumbling infrastructure, dying agriculture, salt-water inundated aquifers, and storms. This small group of men and women are setting a precedent for how society-at-large can relocate thousands from coastal hubs in places as far-flung as Alaska, the Pacific Northwest, the Gulf Coast, the Eastern Shore, and also low-lying coastal lands around the world. Eyes are upon them and lessons learned in Louisiana may even offer clues to South Florida on how to move millions.

I find myself cheering for the members of the Biloxi-Chitimacha-Choctaw Tribe and United Houma Nation. Yes, hidden land mines litter their path, but no doubt their spiritual traditions have ethical codes capable of establishing clear boundaries for decision making. As inconvenient as they will be, ethics can make a critical difference in this high stakes moment. How likely is it that they will turn toward their ethics? I close my eyes and—prayer-like—urge them to resist temptations to gain financial advantage during negotiations, and find ways to behave selflessly for the well-being of all members of the community. I press upon every person in the resettlement effort to listen deeply to what others are saying and not saying and speak only words that they are certain will help. I privately advocate for the men and women of Isle de Jean Charles to embody their brand of ethics in bringing to life a new homeland—one that incorporates the best sustainable energy and green technologies, so the community can live and thrive in ways that offer prosperity, peace, and healing to themselves and to Earth.

I wish the same for residents of Tangier, and I suspect the deeply Christian community of Tangier has it in them to be modern-day Noahs. From what I've seen, residents here have the wherewithal to follow the Old Testament figure's lead in being among the first to adapt to a changing climate by letting go of their homeland and starting anew. Ethics opened the path toward healing once already for them in recent history. It's time for ethics—as a public framework for decision-making—to move center stage again. Tangier residents can lead *themselves* out of difficulty by starting a conversation on how and when to leave, where to go, who will go with them, how they will make a living, and about the ways in which Watermen and their families can continue to access the land and resources of their grandparents. The clock is ticking. By all accounts, they likely have more than six years to relocate, but I doubt they have the thirty years Miami Beach is counting on.

Below is a meditation I wrote to help me apply my ethics when making everyday decisions. It's a spiritual tool that helps me dig deep and see what needs to be done. This meditation is especially useful when the stakes are high or when I'm especially challenged to change my behavior.

> Breathing in, I am aware of the harm I cause using
>> thoughts, words, and actions that kill.
> Breathing out, I aspire not to kill.
> In, killing thoughts, words, and actions.
> Out, aspire not to kill.
> (Ten breaths)

Breathing in, I am aware of the harm I cause when
    I take what doesn't belong to me.
Breathing out, I aspire not to steal.
In, harmful taking from others.
Out, aspire not to steal.
(Ten breaths)

Breathing in, I am aware of the harm I cause when I
    engage in or support sexual misconduct.
Breathing out, I aspire to refrain from sexual misconduct.
In, harmful sexual misconduct.
Out, aspire to refrain.
(Ten breaths)

Breathing in, I am aware of the harm I cause
    when I speak unskillfully.
Breathing out, I aspire to say what is true in
    ways that inspire confidence.
In, the harm of unskillful speech.
Out, true words spoken skillfully.
(Ten breaths)

Breathing in, I am aware of the harm I cause people,
    society, and Earth when I consume unmindfully.
Breathing out, I aspire to moderate my consumption
    and exercise mindfulness in my purchases.
In, harm from unmindful consumption.
Out, consume with care.
(Ten breaths)

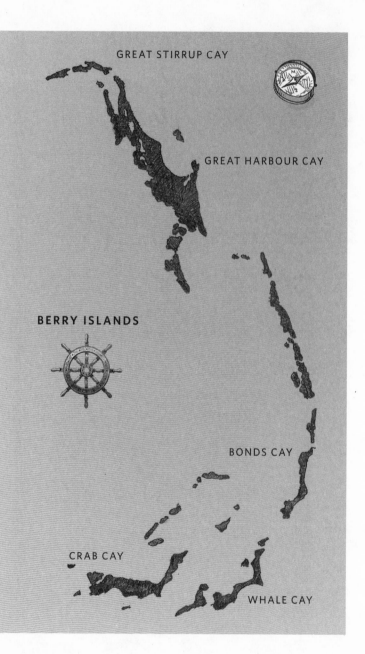

6   TIME

"I love it here," Dave says around a bite of fried egg on toast. Yellow yolk dribbles down his beard.

"Really?" I take in the blue sky set upon blue sea. No land is in sight. Anchoring in open water makes me feel uneasy.

"It's so unusual to be able to stop in the middle of the ocean, drop anchor, and be safe. Where else can you do this?"

"Nowhere I know."

We lingered in Chesapeake Bay until freezing winds and icy rains chased us south. Taking the first right turn out Chesapeake Bay, I was at the helm and on the radio with a Navy destroyer who demanded to know "my intentions," when a single, erratic wave struck *Wild Hair* broadside and nearly knocked Dave off the boat. His fingers snagged the lifeline and he slithered safely on board, but it was a while before he made his way back to the cockpit.

"You all right?" I asked, trying to stay a course in the raucous sea.

"Just a couple of busted ribs." My poor man.

We sailed day and night on the thousand-nautical-mile journey toward warmth. The few times we stopped, we bent our bodies into pretzels to make repairs. Tremors from Hurricane Ida in the Gulf of Mexico stirred the Atlantic, and forced us onto the ICW. In Florida, we spent weeks in a boatyard (again) fixing clogged fuel lines, a busted alternator, fried batteries, and a bent sail track (so we could finally, properly raise and lower the mainsail). It was about this time when my mother began asking if we bought "a lemon." I told her the old mariner's axiom: sailing is moving from one exotic port to another to make repairs.

I was practically giddy the day the fellow who taught us how to rebed port lights and hatches stepped aboard. Buying a satellite phone was another highlight. Between it and the single-side-band radio I now I have two ways to download weather reports or call for help when we're offshore.

Now it's time for more Bahamian adventure. We left Miami two mornings ago, made an uneventful crossing of the Gulf Stream, slept in the lee of a Bimini Island, sailed all day yesterday, and dropped anchor here in the middle of the Bahama Banks. The Milky Way blanketed us through the night; shooting stars stitched the velvet blackness in place. The ocean was gentle, but I couldn't shake the sense of pending misfortune—like when our teenage son was out at night with the car. Now it's daylight and a tropical breeze brings puffs of clouds so close I want to pluck and eat them from the sky.

"You're right. This middle of nowhere is astonishing."

"It's just you and me and the big blue sea, baby." Dave's eyes twinkle.

" . . . And Dinghy. Don't forget her." My eyes move to her perch near the helm but find it empty. "Hey, have you seen her?"

Dave and I scramble in different directions looking first in the water and then behind equipment on deck. The hard fact is if she fell overboard, there's no way for her to climb back on. The hull is too slippery and there's nothing for her nails to grab. Underway, we keep her below or on leash. But in the calm of this place we let our guards down. The thought of a little kitten struggling clenches my gut.

At the back of the boat, on the lowest transom step, I find her content and staring into the water.

"Got her."

"What's she doing?" Dave asks. I peak over her head. A half-dozen minnow float in the shade of the boat.

"She's watching fish!" Dinghy's ears twitch, but her attention stays fixed. "I don't think she needs any more Benadryl." (The vet said children's Benadryl applied to her lips would ward off seasickness. But it was chaos below decks when I touched her mouth with my moistened finger. She spit, hissed, and raced the length of the interior, claws out, tearing upholstery, sending papers and tools flying.)

"Let's cut out that foolishness," Dave says. "She'll do fine now." He unhooks the bungee keeping the mainsail down, preventing it from inadvertently catching the wind and putting *Wild Hair* into motion while we're at anchor. Thrill at the prospect of getting going energizes my limbs.

I munch the last of my oatmeal and blueberries, gulp coffee, stack dishes, and fold the cockpit table down and away.

The shallow waters of the Great Bahama Bank are a rarity. The Bank is a massive, eight-kilometer-high column with an irregular top. It's a pillar that stretches two hundred nautical miles from north to south and east to west. The ocean around the column is fourteen-thousand feet deep, but the plateau surface sits just nine to twenty feet under water—an easy reach for our anchor chain. The highest bumps atop the column comprise the Bahamian islands.

How the column formed is a mystery, controversial among scientists. We do know that five hundred million years ago the single continent of Pangaea split—the Americas parted ways from Europe and Africa—and the Atlantic Ocean flooded the gap. Land on the planet's surface still wanders. Today, the great plates of earth's crust collide and grind each other's edges in a colossal, never-ending train wreck. I learned this firsthand in 1987 when—while living in a suburb of Los Angeles—an earthquake hit at 7:42 a.m. It was October 1st. Eland was one year old and in my arms. I was headed down the flight of stairs in our townhouse, on my way to meet Dave to buy a sailboat (I kid you not) when the Whittier Narrows Earthquake struck. Our dwelling was less than a mile from the quake's point of origin. Mother and child were tossed like lettuce in a salad for twenty-eight seconds—an eternity. The motion was later described by scientists as "complex" (something about a double train of P waves in rapid—less than two second—periods). What's called the peak horizontal acceleration was measured at .63g-units in my locale.[27] I couldn't tell you what all that means exactly, but it strikes me as an accurate description of exactly what I felt, nonetheless. Masonry supports holding the building in place disintegrated. City inspectors declared our home uninhabitable along with 1,470 others in the area. But we were lucky. Eight people died and two hundred were injured when roofs and walls rained down, but ours stayed upright. Plus, the experience shook me awake to the living planet underneath my feet, and shot me back into school to feed a new fascination with all things Earth. (By the way, Dave and I never did buy that boat.)

Not far from where we're now anchored, the North American plate crashes into and under the titanic Caribbean plate. The making of the Bahama Banks started, according to one theory, two-hundred-million years ago, when an underwater volcano at the plate's edge spewed lava not in the shape of a cone but in the shape of a table. The "basement rock" under the Bahama Bank is volcanic. Scientists suspect something else bizarre happened when air currents lifted and transported iron-rich dust from the Saharan Desert to here. They think desert dust fed massive cyanobacteria blooms in what must have been a series of joyous buffets.

During not one but four different ice ages—when the sea level was 340 feet below what it is today—coral reefs flourished. Researchers uncovered countless layers of skeletal deposits from cyanobacteria and coral life perched atop volcanic basement rock. Creatures left their bodies behind—tons of calcium carbonate that we commonly refer to as limestone. While these geologic scenarios remain open to further testing and refinement, one thing is known for sure: the underwater column of the Bahama Bank formed slowly. Column-making began in the cretaceous period, if not earlier.

I'm comforted knowing our geological inheritance: the splitting of Pangaea, four glacial advances, the ongoing battle of tectonic plates. Knowledge of the history of a living Earth helps me make sense of the present—the moments and places where dark, deep ocean transforms into bright, shallow water. Understanding the hows and whys of Earth's contours gives me clues about where the sea is confused and dangerous, where ocean currents meet underwater slopes and well up because they have no other place to go.

It is day three of our journey from the US to the Berry Island Group in the Bahamas. Time glides by like silk on smooth skin. Dave mans the helm as I stand at the bow—with hands on the pulpit as we sail—counting starfish clusters on the ocean floor. Sunbeams slow roast my shoulders, back, and legs. A steady wind draws the boat forward through calm water at six knots.

"What's our depth?" I call.

"Uh," Dave's eyes find the depth gage. "Twelve, thirteen, thirteen." After a moment he adds, "The chart says it's twelve to fifteen feet as far as the eye can see."

What a happy happenstance for our keel five feet below. This is perfect cruising ground. Sails are tight to center because the boat is pointed toward the wind. *Wild Hair* dances underfoot; the ship and I delight together in this day.

"I can see bottom as clearly as I can see you," I shout. Dave smiles and nods across the distance.

I compare the clarity of these waters to the algae-rich, tannin-stained waters of Wisconsin lakes. In my home state, waters are naturally opaque and full of life: grasses, lily pads, cattails, crayfish, blue gill, walleye, bass, musky, and more. Migratory birds birth young in Wisconsin waters because food and nesting supplies are abundant. In contrast, the Bahama Banks appear as sterile as mouthwash. Acres of underwater sand dotted by the occasional algae-basted blade of grass are difficult habitat for most. Only starfish and Dinghy's minnows seem to like it here. But the color of the water is a blue of such bright intensity it hurts my heart. The turquoise is made of three pure ingredients: sand, water, and sun.

I imagine myself a particle tumbling from a cliff off the coast of Africa. Tiny and nearly weightless, I float in the current, lift toward the surface, and am swept into the momentum of a wave; I drift into cold, deep water; I pass through the gills of a fish, snagging on the rough texture of tissue for a moment; I travel inside the length of a whale and emerge unchanged. But when I float atop the Great Bahama Bank, the sea warms, evaporation intensifies, the water grows saltier, and the salinity decreases the solubility of calcium carbonate. Suddenly I am a coated ooid—a particle jacketed in calcium carbonate. Fattened, I settle on the Bahama Bank. In my new home at the bottom of the sea, I am a single grain of oolitic sand glowing under the sun as bright as the calcium carbonate minerals of a vitamin pill. I reflect light as just one of an immeasurable number of grains of sand—miniature moons atop the column—to illuminate water's true color.

The wind gusts and *Wild Hair* responds. Lifting my eyes, I note a dark smudge on the horizon—a building the size of a city block but far off. It is a cruise ship. The floating structure is an odd thing to take in after days of only water and sky. I am moved by a wave of compassion for my species. The ship's unlikely appearance manifests human desire, cooperation, skill, courage, vision, and indulgence, all at once. A sharp stab of light glints from a shiny surface. I make my way hand over hand to the cockpit.

"You see that?"

"Yeah," Dave says. "It's a big one."

I calculate the direction of the ship's movement—cruise ships close their distance in minutes and it's our job to stay clear of them. I determine the vessel must be anchored in place. Then the scruffy low-lying island that is tonight's destination joins the cruise ship in disrupting the horizon's smooth plane.

We will anchor this evening in Bullocks' Harbour on the lee side of Great Harbour Cay. This Cay is the largest island in the Berry Island group. Tomorrow, we will navigate *Wild Hair* through a razor-thin channel up the island's central river to a marina packed with American sport fishing vessels like crayons in a box. Guidebooks promise we will discover a laidback Bahamian settlement with just a few tourists. I look forward to walking about, meeting a few locals, and finding out how life is for them. My mouth waters in anticipation of this year's first plate of fresh, fried conch.

The cruise ship inflates on the horizon and I look for its brand. I have a habit of looking for a particular ship. Nearly thirty years ago—before we were married—Dave sent off for an information packet on how to apply for cruise ship jobs. When it arrived, he typed the template letter multiple times, substituting "I am applying to be a dancer," with "I am applying to be a doctor." It turned out a medical officer on a Norwegian Caribbean Line vessel longed for a vacation from vacationing, so Dave became ship's doctor for the month of March in 1984.

Mid-afternoon sun streams into *Wild Hair*'s cockpit, illuminating the pile of sail lines on the teak floor. Dinghy—wearing a harness tethered to the grab bar at the helm—sleeps on the mound, selling the idea of a catnap. I adjust my seatback, sink lower into the cushion, and let the rhythm of wind on boat consume my body. My eyes slide shut. Dave scratches a pencil on the chart, noting our position.

After a bit, I look again toward the cruise ship. The lettering on the triangular smokestack reads "NCL."

"Hey, that boat is Norwegian Caribbean Line," I say.

"Yeah, I noticed that." Memories of his time as an NCL doctor make Dave smile. "Norwegian Gem," I read. "Wow, that's one fancy paint job!" A garish cascade of polished jewels stretches red, orange, green, blue, and pink from the bow to mid-ship. A thick line of anchor chain points from bow to water, holding the ship in position a quarter mile north of Great Stirrup Cay—an out-island off the northern tip of Great Harbour Cay. A tender ferrying passengers to and from shore zips across my line of sight. Large block print splayed across the hull reads "Bahamarama Mama." I bolt upright.

"There's . . . " my mind spins. "It's Bahamarama Mama! That's . . . it's the out-island. That's . . . " I search for the right word, "that's us!" I look to Dave for confirmation.

Instantly, I'm flooded with stories from Dave's time on the cruise ship. His daily clinic lasted only three hours, so he mostly read detective novels and slurped rum umbrella drinks poolside. The bulk of his patients were not tourists but crew; their busted faces, knife wounds, and steady exchange of gonorrhea were the result of strangers living in close quarters for months at a time. It was the tourists whose maladies made us laugh, then and now.

There was the unconscious passenger, a cab driver in Nassau dropped at the ship's entrance. A large, alarmed crowd gathered. Someone called for the doctor. Dave arrived from the pool in his swimsuit and t-shirt and checked the older woman's pulse, doing what doctors do at such moments. The woman's frightened companion—in a disjointed tangle of words—said one exceptionally helpful utterance: diabetic. Dave sent his assistant running and in short order pressed a syringe of glucose into a vein in the unconscious woman's arm. Within seconds the patient blinked, smiled and stood. The crowd gasped and cheered for the doctor who could raise Lazarus from the dead.

Dave didn't buy another drink for the rest of that tour.

In 1984, I was busy teaching acting and speech at Chicago's Academy High School for Performing Arts. I joined Dave onboard for my week of spring break. Exhausted by three hundred young artists going through puberty together, I was as drawn to a vacation as a cat to a hanging Christmas ornament.

"Dave, it's so good to see you!" I blurted. "I've missed you so much! Wow—you are so tan! And, this ship, it is amazing. Can you help me with this bag? Where are we headed? I'm starving. Can we get food?"

Dave smiled with deep inner calm and put his hands on my hips.

"Hey, baby," his voice slid. "You look hot!"

"Holy cow, you are relaxed." I down-shifted, looked, and barely recognized my man shrouded in tranquility.

"Yeah, baby, it's the Bahamian way. Try it!"

"OK." I giggled. "You make it look really good!"

I adored Dave's formal uniform—dress whites, pressed pleats, brass buttons, and gold-braided shoulder boards. With his medical degree, Dave stepped onto the ship as a high-ranking officer and the uniform was a passkey to anywhere. We wandered deep in the bowels of the vessel to the crew bar for all-you-could-eat lobster and tall piles of flaky bakery tarts. (Who knew crew ate the best food onboard?) Alternatively, we walked past saluting security guards to the ship's command center, the bridge.

"How's it going tonight, Helmsman?" Dave asked with authority.

"Very good, sir," the man in charge said. "It's a clear night and we should have a quiet voyage." The fellow smiled at me and blushed at his own thoughts.

"Excellent!" We found the radar screen and watched its rapid clock hand illuminate specks of islands and boats. All around, dials and gadgets glowed green. We were transfixed, dared not touch a thing. Most incredible was the view out the windows: the ship's bow steaming toward night as daylight slipped behind. Many stories above the surface, it was like riding a comet in orbit: silent, powerful, and fast.

"Carry on then," my boyfriend said.

"Yes sir. Have a good night, sir."

Officer status earned Dave a private cabin with a steward to tidy up and do his laundry. One afternoon after returning from Freeport, I spotted a sexy pair of panties folded neatly on his bed. They were not mine. Jealousy flamed within me.

"Explain this, please," I said, holding skimpy cloth between two fingers as if it were a vomit coated rag.

"Ahh, but . . . I uh . . . Honey, I can't! I don't know whose those are, but I haven't been with anyone else. I swear!"

"Uh huh," I said, unmoved. Anger simmered.

"Really, baby. Let's talk to Francois—he must have switched the laundry. He probably thought they were yours since you just got here!" Speaking auctioneer-fast he added, "The aerobics instructor sleeps with one of the officers. Maybe they're hers. But I never slept with her."

"No?"

"No! I love you!" Pain in his eyes made me feel sincerity in his words.

The next morning, I arrived to my third day of aerobics and the teacher bounced toward me. "Hey, sorry about my underwear showing up in Dave's room. Don't worry, we've never done it! I'm with Doug."

"Yeah, well that's good." I forced a laugh. "I'm glad to hear that because it looked really bad for Dave!" I noticed the good looks of the woman before me: firm body, thick hair, and glowing tan.

"No really, Dave is a nice guy. You've got nothing to worry about!" She bounced to the front of the room.

She was right: Dave was and is a nice guy: faithful, honest, generous, protective, even-tempered, funny, and loving. He is my lifelong lover, companion, and best friend.

Dave turns his head from the scene to look at me, speechless. Sailing upon the cruise ship illuminated, in each of us, thirty years of both remarkable and unremarkable moments—births, deaths, jobs, houses, intimacies, successes, and failures to erupt. Full-color experiences—complete with smells, sounds, and texture—splay in no particular order like a deck of cards bent between strong fingers. We are standing side-by-side in awe of the massive oak tree (with perfect half-moon canopy) behind our first house; I hand Dave our little daughter as she wails over a newly broken toe; and we claw with bare hands in clay soil to expose wooden triceratops bones Dave and I built from wood and buried for a Jurassic-Park-style birthday party. Thanksgiving turkeys and Christmas gooses on large platters move from my hands to his; music drowns our laughter as knees bump during ballroom dance lessons; I can hear the murmur of pine needles underneath our feet as we hike together. Our current Midwestern home becomes our first city apartment. Faded vacation photos resaturate until we are again on our honeymoon in Cancun. Our son in his prom tuxedo turns into the child in a onesie making crashing noises as matchbox cars slam. With every Technicolor memory, another intercedes.

The sensory parade includes Dave's brother before he died: a cigarette hanging from his lips and eyes twinkling, laughing over the morning newspaper. I see Dave at my parent's dining table, laboring over medical texts. The slight form of my mother-in-law losing her battle with lymphoma is replaced by the soft ground of our favorite campsite. A loon calling at dusk. A cool breeze lifts goose bumps to the surface of our skin. The gentle love we made this morning is devoured by the hot passion of a roadside quickie in Kentucky, and then by our first kiss leaning against the hood of Dave's new car.

The memories continue. I'm crying from the sweet blend of ice cream and parental satisfaction as bike helmets bob and Dave and our young teens jabber about the day at the neighborhood bait shop. Nervously, I walk with Dave into an architect's paneled office to begin the process of designing a new home. Lonely and restless, I pace and stare at the Chicago skyline as Dave navigates the Alaskan Highway without me to a temporary job in Anchorage. The whiplash of emotions is too much.

My eyes find Dave's. He too is undone. I move to him and we laugh and cry in the swell of our shared life. Stumbling tender-hearted into the crossroads of personal history, our young and old selves collide like tectonic plates. Nothing is faded by the distance of the years.

Still, figures mingle, time jumbles, and images unfold. The pool party in California is interrupted by a stadium of shouting Wisconsin football fans. The Mothers' Day brunch at "The Club" in Tucson with Grandmother appears. An argument with Dave's father on our first sailboat over whose turn it is to be captain merges with shouts from regatta racers to "get the hell out of the way!" Bodies of water flow past: the noisy curl of a Pacific wave, the chop of the Mediterranean Sea, the icy chill of Lake Superior, the glaring sun reflecting off the Zanzibar Channel, the gray steel of Lake Michigan, the dazzling blue of the Bahama Banks.

Hurts surface too. Dave stands next to the bed I sit on after I've slashed him with unkind words; I know he's wounded because his pain reverberates in my body. Then I am folding forward—touching my forehead to ground while straddling a meditation cushion. Hot tears drip from the tip of my nose as teeth grind. Our union is tested. But the pain we've inflicted hasn't derailed our journey; it has only made us more careful and deepened mutual understanding.

I see a pattern in these memories. Simple expressions of love between us are my strength and my substance. I am because we are. The whole of my life is contained in our union.

I pull back and look Dave in the eyes. His hand touches my cheek, dries my tears. Laughing, I lift the tail of his *Wild Hair* t-shirt and tenderly mop his cheeks and beard.

"Oh my," I say.

"Yowza."

"Did you have ... memories?"

"A bunch of 'em, yeah. Still do." He neatens a strand of my hair.

"It was the Bahamarama Mama that did it—like the picture," I say. For years Dave carried a photo of me on that beach with the ship's tender in the background.

"You're still my Bahamarama Mama, baby." His hand squeezes my knee and gratitude swells in my chest. At ages 50 and 56, I am still charmed by his adoration. Dave's wrist relaxes again on the wheel. His eyes are steady, lit by joy. His body moves with the motion of waves. *Wild Hair* hums in the wind as azure waters caress the hull.

Yes, the years rest on Dave's shoulders. Events score his face. We are both sagging, graying, and setting unbalanced chemistries right with prescriptions. But Dave remains lean and strong. My fingers reach to the prickle of his short-cropped beard and I lean forward, delivering a kiss of passion.

I know it's not easy being Dave. He's so bright, his mind never stops. He mulls circumstances, identifies potential problems, and—as expressions of love—suggests ways people can avoid trouble before others even sense a threat. Analyzing and reanalyzing situations made him an excellent father, surgeon, and now sailor. But some who would rather throw caution to the wind can misunderstand. His gift at spotting future trouble compels our children to call him "David-Downer." Others who prefer a carefree approach find his contributions meddlesome.

But Dave has the most compassionate heart I've ever known. In his twenties he dropped from a general surgery residency not because it was difficult but because patients sometimes died. It tore him up. Instead, he devoted his talents to helping and healing disabled children because kids want so badly to get better.

I see Dave leaving the house. A thermal mug balancing three slices of rye-peanut butter toast is in one hand; the other holds keys and a scuffed briefcase. He kisses me goodbye at dawn. I make out three-inch cartoon bugs on his silk tie. The rustle of covers gives way to the rustle of winter jackets as the family trundles through snow in hunt for the perfect Christmas tree. Of course, Dave wants the bent and sparsely limbed spruce because no one else will bring it home.

Now we are beach-walking in Baja, Mexico, far from the vacation cottage. We decide on a short-cut—through the field of cows. While walking, we realize we're not among cows but bulls. A hoof scrapes dry ground like a gravedigger's spade. The two-thousand-pound fellow jerks his head up and down, flashing the whites of his eyes. Dave grabs a rock and starts jogging away down the road. I grab a beer bottle and move to catch up.

Once we make it to safety, I ask, "Hey baby, were you going to leave me behind with those angry bulls?"

"If I had to."

"Wait . . . what? Wouldn't you try to protect me? You know I don't run as fast as you!"

"Right, you don't run as fast as me. So of the two of us you were going to get gored."

"Hold on . . . . You were going to dash off and let me die?"

"No honey, I was going to come back and patch you up. If I got gored could you have patched me?" I growl with dissatisfaction. "Don't worry. baby. I love you. Let me take care of you by taking care of me first."

Time has proven this to be an effective strategy. The man's self-preservation instinct is profoundly practical and weirdly comforting. I adjust my seat in the cockpit to keep my foot from falling asleep. My body nestles into the safety of his arms. Dinghy stands, stretches, and steps a rung at a time, like an old woman, down the companionway to her food bowl.

In my mind's eye, I see us readying the boat for departure in silent harmony, deciphering life's events over dinner, falling into sleep with one person's hand against the other's shoulder and our feet touching. This is what is and always will be holy between us: gestures of cooperation, shared learning, and unity.

My imagination expands and images of past relatives surface. My nose appears in a family photo album on my father's, grandmother's, and great grandmother's faces. My eyes radiate from my mother's graduation portrait. I am the continuation of the bodies, labors, and hopes of my ancestors. Silently, a small well of confidence opens in my core.

I am also a product of America's culture. Standing inside a voting booth, my hand grips the black marker and I scribble the blank for my preferred candidate. I sit curbside watching our kids march in the

Memorial Day parade. I search desperately for a parking place at the mall two days before Christmas.

Feeling warm, I stand in the cockpit to cool the backs of my legs. This elicits an image of me flapping fins and tail—dragging my belly through mud—to slither from the primordial soup. No doubt, I have legs today because bony, strong fins were useful in my evolution. Dave was likely with me when I crawled into daylight that first time. Was he faster than me then, too? Was it his day to go first or mine?

The walking-fish daydream cuts to modern me driving my car. Driving is possible because when I was a plant—before decomposing organisms evolved—my dead body fermented into fossil fuel. Perhaps Dave embraced me as the lava that sealed my fate.

Now I am the speck of iron-rich dust from the Saharan Desert feeding Dave—the cyanobacteria. Joyfully, I give him life and delight in his satisfaction.

At age twenty-two, standing apart in a crowded restaurant, my eyes connect with Dave's for the first time. Before rational thought, I recognize him as my dearest friend. Some may call this love at first sight. But seeing Dave across the room feels like coming home. We are mysteriously preimprinted upon each other. There is relief in the remeeting. Later, as I learn the details of who he is, there is the sensation of relaxing into what is already known. I rediscover my eternal partner, mentor, student, and foil.

Aboard *Wild Hair*, I caress Dave's permanently impermanent hand. If the past powerfully shapes the experience of the present, does the future equally mold the here and now? Concerned for our children's future, don't Dave and I insist they go to college? Didn't potential mishaps at sea inspire us to install the single sideband radio, satellite phone, and life raft? Ghosts of both past and future events mingle in the present to inform our decision-making.

The implication of this gives me an electric jolt and I inhale. I have touched infinity here at Great Harbour Cay, revealing the limitations of my concept of time. Before, I shrunk my life happenings into a series of linear events. But these infinite scenes from my life are clearly not linear. The whole of it exists at once. Reality, it seems, lies beyond our ideas of time just as the ocean lies beyond the wave and the atmosphere lies beyond the cloud.

Perhaps Buddha was seeing the power of infinity when he told his disciples to drive all moments into one. He wanted people to see the present moment not as a fleeting way station on the plodding road of time, but as a gateway into both the wisdom of generations of human life and the universal consciousness of the cosmos.

On any other day, I might have dismissed the Buddha's instruction to drive all moments into one as simply poetic. But now I understand the directive is literal and of vital importance. This moment contains all the elements of infinity and the wisdom of the whole. If all of time is in this moment, when we transform this moment we profoundly alter both the past and the future. My eyes scan the horizon as I absorb the possibility and promise of now. My confidence builds. I breathe deeply and with ease.

At the moment of Buddha's enlightenment, he sat under a bodhi tree and vowed not to rise until he achieved insight. As hours passed, doubt flooded him in the form of Mara, the Tempter.

"Who do you think you are, you who are so far from perfection?" Mara said. "You—a little, meager man—cannot possibly become enlightened because *I* am the enlightened one! My armies bear witness to my power. Who bears witness to yours?"

The man slid his hand forward until finger tips touched Earth. With this simple gesture—the story goes—the planet roared in witness to the Buddha's many lives and pure aspiration for enlightenment; the reality of all phenomena throughout space-time arrived to the present to shake and transform the universe.

Mara fled. By morning, the Buddha was fully and forever enlightened.

When we touch infinity, we know unequivocally who we are. We see ourselves as the continuation of ancestors and the forerunners of future beings. Past and future generations come alive and guide our actions. Suddenly, we are part of the ongoing story of evolution, integral to life's continuation, with the power to behave in support of the whole. The meaning of life becomes clear: we are simply to love and be loved.

Today the Great Atlantic Teacher taught me the value of lifting my gaze to the whole of space-time—to a transcendent vision that includes the past, present, and future. Up until now, I was tricked by everyday events into believing time was only linear. I spent my life caught in the sequential details of business and family life. My priorities—do this and gain that—were childlike; they lacked a timeless perspective and universal view. But now, meeting my young self in the wilderness of the ocean, I gain the insight of infinity.

I feel light knowing that this instant is the culmination of a great inheritance: the collective wisdom and workings of humans and non-humans handed down in a tide of universal love. But it's more than that. This present moment is a pivot point for me to pause and regroup before springboarding into the future.

～～

Peace settles down my spine with the grace of a falling flower petal. Dave turns the wheel hand over hand and *Wild Hair* spins right.

"Let's drop the sails," he says. "We've arrived."

REFLECTIONS ON TIME

I've loved Dave my entire adult life, but sailing upon our young selves from the perspective of seasoned maturity lifted the whole of our story into an abrupt, life-altering consciousness. All of a sudden, the fabric of our intimacy showed—its length, strength, elasticity, and texture, its function as the organizing element of my existence, as the well-spring of joy nourishing my days. At Great Harbour Cay, I woke up to a larger reality—our unique manifestation of love I'll call "us," and the effect of us on others. Now, I see our sacred universe, all it contains, and I can observe how we spin through a galaxy of love born by others. Time itself broke free from the shackles of my linear conceptualization and the infinity of us sprang forth—more alive and palpable than before. It burned through my narrow misperception of our existence to reveal the boundless magnificence of our relatedness.

When I look deeply into the mystic quality of the experience, a question bubbles up: "Heather, what the hell else did you think was going on all this time?" Honestly, I don't know. The dearness and meaning of our union was never hidden, but I was a fish swimming in water, too close to comprehend the ocean's brilliance. I suppose day to day details and a never-ending string of deadlines kept me from driving all moments into one, deeply taking stock, reflecting on the full worth of our bond and its meaning to my life. It took the Great Atlantic Teacher to set loose my rapid shift in consciousness. And when I think about it, I can't consider what happened to be any kind of a supernatural encounter—the experience came from nature itself. Infinity must be available to each of us, always.

It's no wonder that on the eve of Buddha's enlightenment, when he put fingertips to ground so the planet could witness his many lives, Earth roared. To my mind, the reverberation was infinity arriving in an instant, and it transformed *everything*. In that moment, the Buddha stopped suckling the teat of society's thin milk and drank from a deeper spiritual well. Inner knowing arrived when his relationship with the whole of space-time lay bare. And I find it especially beautiful, too, that he asked Earth to witness his awakening. The asking underscores Earth's relationship to humankind as a beloved mother and ally. I think he understood Earth as the leading character in his personal story and in the story of human history, appreciated the countless ways in which Earth's elements and processes sustained his life, and knew the cycle would continue for eons after the last human's final footstep. Earth is here all the time, and the *whole* of it is present now.

Earth is humankind's unblinking witness. The stories of the planet's making, the evolution of life, and human achievement can of course be taken separately, but they are also one *epic* tale. Peoples throughout time have tried to make sense of the world through stories, to describe their place in the world. But in modern times, we split consciousness, parse events, and create a gulf between the natural world and the human realm. Many of us hunger for a fuller understanding of our place in space-time, and authors like Brian Swimme and Thomas Berry respond with a new story of the universe, one that celebrates humanity's place in the cosmos and the experience of time itself as something they describe as *the evolution of "irreversible transformations."*[28]

Of course, linear time is essential to daily life. We must show up to the dentist's office at the agreed upon moment, or else. But when we bring infinity into our relationships, we become conscious of the *fullness* of our intimacies, appreciate how precious our beloveds are, and steward our connections with care. When we bring infinity into our relationship to Earth, we invite the workings of evolution into our consciousness. Suddenly, our days are not plodding but *miraculous*. We become aware of our shared genetic heritage with other animals, and our existence is suddenly and without warning knitted into the continuum of life. We feel the energy ricocheting in our body as something different—a vitality lent to us from the force released at the moment of the big bang. In this way we appreciate the energy of the sun, wind, waves, earthquakes, and plants as the very thing animating *our* bodies. Opening to the insight of infinity, the human story becomes fully assimilated with that of the mountain, the moon, and the starfish.

Driving all moments into one, we know who we are in the universe, understand what is beloved, feel humbled and grateful to human and nonhuman kin, and act respectfully. When the infinity of time is alive and palpable, we know how to be in a suffering world.

Sometimes I think climate disruption is born of recent events. But when I look through the lens of infinity, I can see the climate change story includes a trio of ancient hunter-gatherers choosing not to wander but to plant seeds—to cultivate and harness the environment in service to humans. Their decision launches the Neolithic Revolution. In our more recent history, Matthew Boulton and James Watt huddle near a window in 1775 reviewing plans for a coal-powered steam engine, thereby giving birth to the Industrial Revolution. I see Karl Benz in Germany, on a chilly New Year's Eve in 1879, starting the first stationary gasoline engine—what would become an essential component of the automobile. In 1970, Brazil's military dictator General Medici holds the hand of a starving child in his country's drought-stricken North-East territory and decides in that moment to build a 3,100-mile-road to occupy the Amazon Rainforest, resettling ten million people onto small farms and bringing catastrophic loss to Earth's last expansive virgin forest. Infinity erases all notions of fault or blame. No one knew we'd end up here. And it's impossible for anyone to say for sure where humanity is going.

That's the thing about infinity: the story has no end. Earth will go on, the cosmos will continue. But there are chapters in the story of

Earth—sequences of transformation—and how we live in *this* moment will determine what comes to be in the future.

As an individual, I want to keep my consciousness awake by devoting at least part of my awareness toward the insight of infinity. I want to maintain gratitude and reverence toward my relationships (with people and with Earth) and to avoid getting swept into things unimportant or inconsequential. I do this because the act of cultivating *understanding* is climate *action*.

I can bring infinity into consciousness through simple practices. I keep my ancestors alive by speaking with elders about their childhood, learning my family's stories, and celebrating our traditions. I can see myself at this moment in infinity when I look in a mirror and ask, "Eyes, where did you come from and where are you going?" I can connect with future generations by role-playing with a friend pretending to be a "future being," taking turns sharing what's in our hearts. I can write a note to someone who will be born one hundred years from today and seal it in a time capsule. Or, I can sit and walk in nature and contemplate my interrelatedness with Earth—my witness, my ally, my mother.

When infinity yawns wide, I stand with thousands of people across time and space in my aspiration to heal and protect Earth. Together, we advocate for planetary peace and well-being. Driving all moments into one, alone-ness evaporates and together we move in solidarity as a single, massive expression of love.

Below is a pleasant meditation I like to do to drive all moments into one and experience the insight of infinity.

> Breathing in, looking with eyes of infinity,
>> I see myself bound to my beloved.
> Breathing out, I smile in gratitude to the fabric of our relatedness.
> In, bound to my beloved.
> Out, smile to the fabric our relatedness.
> (Ten breaths)

Breathing in, I see myself alive before I was born in
    the body of my great, great, grandmother.
Breathing out, I see myself alive before I was born in
    the body of my great, great, grandfather.
In, alive in great, great grandma.
Out, alive in great, great grandpa.
(Ten breaths)

Breathing in, I see my actions continuing 100
    years in the body of society.
Breathing out, I grow mindful that helpful and
    unhelpful actions will live on.
In, my actions continue.
Out, mindful of my actions.
(Ten breaths)

Breathing in, I see the energy of the big bang animating my body.
Breathing out, I smile in solidarity with the universe.
In, energy of big bang.
Out, solidarity with universe.
(Ten breaths)

Breathing in, I trace the saltwater of Earth's oceans through
    generations of evolution and into my tears.
Breathing out, I smile at the insight that I am Earth's oceans walking.
In, oceans in my body.
Out, I am Earth's oceans walking.
(Ten breaths)

Breathing in, I see Earth herself as my
    witness, my ally, and my mother.
Breathing out, I am grateful to Earth for being
    my witness, ally, and mother.
In, Earth as witness, ally, and mother.
Out, grateful to Earth.
(Ten breaths)

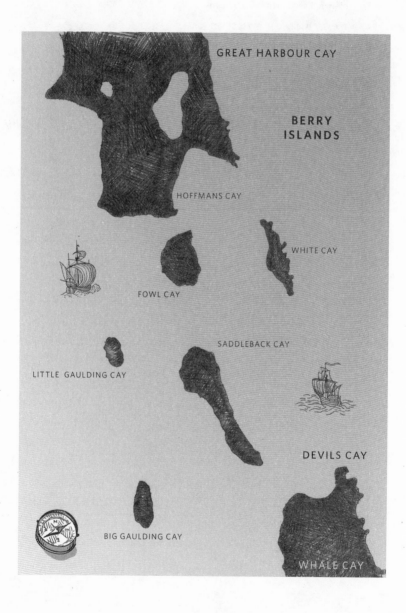

GREAT HARBOUR CAY

BERRY ISLANDS

HOFFMANS CAY

WHITE CAY

FOWL CAY

SADDLEBACK CAY

LITTLE GAULDING CAY

DEVILS CAY

BIG GAULDING CAY

WHALE CAY

7 POWER

"Mayday, Mayday, Mayday. This is sailing vessel *Wild Hair*, *Wild Hair*, *Wild Hair*. My position is north 25 degrees 36 minutes, west 77 degrees 43 minutes. Repeat: north 25 degrees 36 minutes, west 77 degrees 43 minutes." I swallow hard; my spit feels like sand in my mouth. The tempest screams. I fear Dave is washed away.

"We are west of White Cay in the Berry Islands of the Bahamas. Our vessel is on the rocks. One person is on board the vessel, and one person has blown out to sea in an inflatable dinghy. Mayday, Mayday, Mayday." I wait for a response. The radio crackles into the void. "SLOWLY—CLEARLY—CALMLY," the placard on the radio reads. "Follow these instructions ONLY when there is IMMEDIATE danger of loss to life or property."

No problem there. After a long beat, I repeat the Mayday and location but feel incapable of explaining the exact nature of my distress.

"I am onboard the vessel but the captain has blown out to sea—into the Northwest Providence Channel—in a ten-foot dinghy. *Wild Hair* is standing by on VHF channel sixteen and SSB distress channel 2.1820. Please, come back."

In my imagination I see my husband wearing a headlamp, raincoat, and an inflatable floatation device. The outboard is dead and there is no paddle. All he can do is hold on as the storm flings the dinghy back and forth in the thirteen-foot seas roaring outside the harbor. The image makes me dizzy with helplessness and I slide into the booth. The boat, resting on the ocean floor in a falling tide, tips off-kilter. I use my feet to wedge in place.

The EPIRB, our ship's Emergency Position Indicator Locator Beacon, glows neon yellow with potential in its mount. Maybe I should flip the switch, bring a search and rescue team here to the radio transmitter. But I don't want coast guard helicopters here because I'm perfectly safe aground in the boat; I want the Jayhawk and Falcon aircraft—with full military force—hovering in the Channel searching for Dave. Confusion is the last thing we need. I've got to talk to someone so they understand.

I send a third Mayday. Static sizzles in response. I feel impotent, unable to help Dave. "C'mon, Heather. Keep it together," I say aloud. My will keeps my emotions in check. I take a breath.

The VHF broadcasts as far as a person can see given the curvature of the earth. It's a line-of-sighe radio. White Cay is uninhabited, but I'm hoping sailors in the one boat anchored a half-mile away have their radio on and that they have the stuff to come out tonight. I keep the VHF tuned to hailing and distress channel sixteen.

The SSB is a marine version of an amateur ham radio. Its signal has horsepower enough to travel to the ionosphere and back to earth at various angles and distances, depending on the frequency. I dial the second of four distress channels—4.1250. Perhaps someone a little farther away will hear my cry.

Suddenly, I remember the satellite phone in the cupboard. I preset seven Coast Guard emergency phone numbers for stations up and down the East Coast to use in a pickle. I lift the phone from the case. It is surprisingly heavy. My finger hovers above the buttons.

"OK, Heather, how do you find the preset numbers?" I blink and scan the surface for clues. I close my eyes and try to remember. *Wild Hair's* hull scrapes rock as it tilts. Sweat beads on my nose.

"Arrr . . . I can't remember! You're nothing but a paperweight," I say, dropping the phone in its case and taking charge of the radio mics again.

"Mayday, Mayday, Mayday, this is sailing vessel *Wild Hair*, *Wild Hair*, *Wild Hair*." I repeat the script three more times, pausing to listen and breathe in between. Nothing but white noise leaks through the speakers.

The DVD cover for the TV series *Lost* falls from the sofa and slides across the tipped salon floor. We chose this harbor days ago because the weatherman predicted back-to-back storms, and we thought we could find protection within this circle of islands. One storm passed harmlessly. Thinking we had things under control, we ate a big dinner and started binge-watching season three when—simultaneously—the tide reversed and the storm clocked, unhooking our anchor from the sea floor.

Bam-ba-bam-BAM. The hull heaved and shuttered to a stop. Dinghy, who had been asleep on my lap, sprinted like an Olympian to the forward berth and hid behind a spare sail.

"What the hell?" Dave said, leaping the stairs to look outside. "Heather, start the engine—we're aground!" I flipped the switch to arm the controls at the helm, and Dave started the motor and put it in gear. The engine revved, but rocks blocked *Wild Hair*'s keel. Our attempts to drive off were useless.

A pen and a few papers follow the DVD's lead onto the floor. I question my safety as the ship tips further. "Well, Heather," I reason, "The tide is going down, so you're going to keep tipping. But the boat can only slant to the side so far. When the tide returns, the boat will right. If there's a hole in the boat . . . well . . . the boat is far enough on shore that it won't fully submerge. You're safe, honey. Focus on Dave." Fresh panic pricks my muscles. I flip the SSB to the third distress channel: 6.2150.

"Mayday, Mayday, Mayday . . . ." After two more tries, I switch to SSB channel 8.2910—the final distress channel. Still, I'm met with nothing but silence.

"God damn that outboard!" I set the mics down and rest my head on my balled fists. Anger surges through my limbs. The bloody thing started acting up months ago in the Chesapeake, and has only worked intermittently ever since. Experts told us the ethanol additive in gasoline had coated the motor's interior like varnish. We carried it to a shop where a mechanic disassembled, steam-cleaned, and reassembled each part with the precision of a Swiss clockmaker. The outboard still doesn't have the goods, despite a strict diet of marine-grade gas. Every other day, Dave studies the manual, surfs discussion forums online, or speaks with someone at the manufacturer's technical support hotline. We keep thinking the problem is fixed, but then the beast dies again. It's been a pain, but I never imagined the problem could prove fatal.

"Run through the SSB channels again, Heather." I double-check the VHF is on sixteen, dial the SSB to 2.1820, and start the script again. SLOWLY—CLEARLY—CALMLY. Dave's life depends upon me. I recommit to the task with vigor.

A dragged anchor isn't always catastrophic; sailors muscle their boats to safety by setting a spare anchor ninety degrees from the hull and—as the tide rises—winching the ship in the same way a tow truck hauls an automobile from a ditch. Our self-rescue drill began smoothly enough. I got the spare anchor from the lazarette and passed it and the line to Dave in the dinghy. The small inflatable boat made wet rubber balloon sounds against the hull as it bounced in the three-foot waves. Dave yanked the starter cord, the outboard sprang to life, and my man zoomed away from me. An unlucky wave knocked the anchor from the dinghy gunnel too soon into the drink. Then the outboard choked and quit. The dinghy, carrying Dave, slipped sideways in the wind faster than a car on ice.

"FUCK!" I could hear Dave's voice from the boat.

The moon was full, the light sharp. Events unfolded in black and white, with the high drama of a silent motion picture. Dave struggled to restart the outboard. *Rip . . . rip . . . rip.* His headlamp swung wildly. The 35-knot winds pushed the inflatable and Dave slipped through an opening between two islands and was gone. I stood in shock, unable to breathe.

The sea beyond was a violent expanse. The next landfall is continental Africa; the whole of the Atlantic lies between. I was alone, naked to nature and its dominance.

Gusts pushed me about but didn't knock me down. My hands took hold of stations and lifelines. My eyes remained frozen on the gap, confused. "Should I issue a Mayday?" I talked to myself for comfort. "No, Dave is a cat with nine lives who always has a trick up his sleeve." This sounded ridiculous. "Yes, of course I should. He's going to die." I rubbed my hands upon my face to stimulate inner strength and decided to count to ten—give myself time to think.

"One . . . two . . . ." Dave's last word echoed down my spine and stripped my confidence. "Three . . . four . . . ." I recalled the equipment stored in the dinghy, then realized the storage box was at my right foot on *Wild Hair*'s deck. He lacked a radio, EPIRB, motor, or paddle. "Five . . . six . . . ." I looked toward roiling cumulonimbus clouds. Dense towers stretched to twenty-thousand feet. They undulated like dragons in the wind. I felt defenseless—like a prisoner before a firing squad. Even so, the cloud's beauty surprised me.

"Seven . . . " Then something really odd happened; it was something beyond words. A great causal stream registered with me, deep down: this was because that was, and that was because of this. In the constantly flowing river of cause and effect, there was nothing and no one to blame.

"Eight . . . " Again, the wind pressed my rain jacket and separated around my body. The wind accommodated my presence, and this struck me as profound. It accepted I was there and adjusted its course without complaint. The universe allowed me to be. I concluded from this gesture that there was nothing personal in *Wild Hair*'s twist of misfortune.

"Nine . . . " I looked west toward the storm and took it all in—the magnificence of the blameless, untamed wilderness. I looked east toward the island and the gap; my longing for Dave felt knife-sharp. I looked down at *Wild Hair* and noted the starboard deck was low. I looked up at the moon and witnessed the purity of light reflected. In the whole of my view, I was the only thing with consciousness. Each element in the world around me responded to stimuli without thought. Everything moved in simple cause-effect relationships. I alone had the power of choice to decide how I would respond to events.

"Ten . . . Heather, call a Mayday."

~~~~

I bump the SSB dial to the next frequency—4.1250. "Mayday, Mayday, Mayday." I keep struggling in every call to articulate the nature of the emergency in a single sentence. I wait, listening to the nothingness of space. I roll forward to put my forehead on the desk. My muscles clamp tight. I feel completely helpless. "Maybe I should trigger the EPIRB beacon after all," I whisper. "Where is everyone?"

"Hea-THURR!" Dave's voice shouts above the gale.

I spring upright. Thoughts jumble. He's here! Where? It's my imagination. Change frequency and call again. No—stop! Put the mics down. What if someone answers and I miss them? Move!

I climb the tilted ladder, finding new handholds to stay upright. My pupils adjust. *Wild Hair*'s mast points oddly to shore. I search for a swimmer in the water, but spot Dave slide-walking down the island's steep dune. Sand avalanches with each step. He moves stiffly—arms extended—down the forty-foot face. Arriving at the beach, he trudges through calf-high water toward me. *Wild Hair* kneels for him like a trained elephant, and he easily hoists himself onboard. I step to him. My man is drenched, shivering, and unnaturally large because the personal floatation device triggered beneath his rain jacket. The swelled inner tube under his clothing doubles his normal size and prevents him from embracing me. I wrap his shaking, bloated form in my arms, overcoming the fear locked in my muscles.

"Damn outboard—"

"How did you get here?"

"I walked."

"I saw you blow between the islands . . . out to sea."

"No, I didn't. I paddled to the island and walked home. I've got to go get the dinghy and we've got to try to drop the anchor again."

"Wait, you had a paddle? I've been calling a Mayday." Dave looks at me. "I thought you were out there!"

"Oh boy, no. Didn't you see me? I paddled like hell to shore. I guess I caught the backside of the island. There was a shoal. It was underwater when we came in, but it's exposed now."

"—past my sight, where I couldn't see you?"

"I guess." He takes in my fear. "Sorry, baby."

"You're wet."

. "Yeah. I waded the shallows and there was a hole. I was walking, then I was swimming." He laughs. "Boy was I surprised! It blew my life jacket. We'll have to get another cartridge to arm it."

"Forget that, you're here!" My hands grip his forearms, hard, and the capacity to smile returns.

"Yeah, I've got to get some warm clothes on. Then I'm going back to get the dinghy and we're going to try this again."

"No! We are absolutely not doing this again." My determination is ferocious. "I just spent the last however long calling a Mayday because I thought you were battling for your life!" Frustration crosses his brow. "We will not trust the outboard again with your life. No way."

"Heather," his words have force. "I am not willing to lose this boat! When the tide rises the wind will push the boat higher onto the rocks. We'll never get it out then. We're lucky this happened at low tide. We have a chance to save her. We've got to try again."

"Not with the dinghy we're not. Absolutely not. I'll lose the boat. I'm not losing you." A cloud passes in front of the moon and the world darkens.

"OK, look." Dave is pissed. "I'm going to change clothes because my teeth are chattering and I've got to conserve my energy. It's going to be a long night. Get the spotlight and send an SOS to the boat across the harbor. Maybe they can help us with their dinghy."

I pull the spotlight trigger and the beam soars through sheets of rain to illuminate the neighboring ship. *Short-short-short-long-long-long-short-short-short. Pause.* I repeat the series of short and long SOS flashes, wait several seconds, and repeat them again. And again.

"C'mon, c'mon. Don't be asleep. We need you!" I check my watch: 10:17. Hopefully, our neighbors are still awake. The blinking beacon shines bright on their boat. *Short-short-short-long-long-long-short-short-short.* Far off, a light switches on. A figure, silhouetted by the moon, climbs outside. The tip of a cigarette glows orange in the distance.

"Yes, I mean you!" I flash the SOS repeatedly. After a beat, the figure goes below. The light turns off.

"What? Buddy, you have got to be kidding me. Get over here!" My finger grows emphatic on the trigger. *Short-short-short-long-long-long-short-short-short.* Over and over I signal. The figure and cigarette reappear. "Yes, that's right. You *come* here. You *come* here. You *come* here," I say in synchronicity with the beam.

Again, the figure goes below.

Is he getting dressed? Mobilizing for a rescue? He should give me some sort of sign that he's on his way. He is certainly taking a while. The neighboring ship's inside light flips off.

"No, no, no, you son-of-a-bitch! I will haunt you with this light all night if I have to." My pace becomes frantic. My teeth clench. *Short-short-short-long-long-long-short-short-short. Short-short-short-long-long-long-short-short-short. Short-short-short-long-long-long-short-short-short.* Minutes pass. The light flips on. The man steps into the moonlight again.

"OK, you motherfucker. Look at this!" I swing the spotlight toward the island and set a wall of sand and stone ablaze. I scare myself. The nearness of the mass shatters nerves. We are *on* the island. With longer arms, I could touch the wall from here. Tree limbs twist in the glare—a haunted spectacle. I swing the spotlight back to my harbor-mate. "Got it?" I ask with my short-long-short voice.

The figure moves at a faster pace. Back and forth it walks. It disappears quickly below and reemerges. I light the man's actions with my beam so he doesn't stray from his purpose. *Short-short-short-long-long-long-short-short-short.* Soon, the man and his dinghy steam toward us through the sideways wind and rain.

He is a chiseled Frenchman—slight but able-bodied—with not a word of English. I hug him as he steps aboard. He smiles.

"What is your name?" I ask above the wind.

"A-gggha—" he replies.

"What? I didn't quite hear you."

"A-GGGHAN."

"A-GGGHAN?" I question.

He nods and smiles. "AGGGHAN."

With hand gestures and a smattering of universal sailing terms, a plan develops. The men will tow our dinghy home. Then, using Aggghan's dinghy, the two will set the spare anchor in place.

"Wait!" I shout before they depart. "Aggghan, you need a PFD."

"*Non, non, non—*"

"Yes, yes, yes," I say with force. "This is a night of DANGER!" I hand him a spare PFD with a look that means business.

When the men return, I tie our dinghy to the side. Powered by Aggghan's outboard, Dave and the Frenchman drop the kedge-anchor ninety-degrees from *Wild Hair*'s hull; I loop the anchor rode around the winch and turn the handle until *Wild Hair* is under tension. Back on board, Aggghan removes the PFD, which in the fracas spontaneously inflated.

"So-ree," he says.

"No, no—it's good," I hug him again. "Thank you!" Dave shakes Aggghan's hand. Our helper retreats to his ship.

~~~

It is low tide. The boat is tipped over, the mattress is vertical, and I am essentially standing sideways across it with feet wedged into the wall. I can't sleep. Dave snores loudly from the only real bed available with the boat on its side like this—he's snoozing on what normally is the backrest of the salon sofa. His comfort makes me *insanely* jealous and I want to rattle him awake and rant about his selfishness, but I don't. We'll need his well-rested brawn in the hours ahead. Time passes.

A loud ripping sound rakes my attention. Its 05:04, the tide is rising, and *Wild Hair* is starting to float again. But with each surging wave, the rudder is slamming backwards onto a submerged rock. I get up and walk along the wall to get out of bed and grab my shoes.

"Let's drive her off, now," I say to a groggy Dave as I wiggle a heel into place. I arm the engine, climb the tilted ladder into the cockpit and slide behind the steering wheel as he follows me and gets busy attaching a plastic float to an anchor line.

"I'm gonna release this line and anchor as we drive off," Dave explains. "The float will tell us where it is so we can pick it up later. " I nod as I hand-over-hand the wheel hard to port and lever the throttle to full. The engine roars and the boat tugs. The hull inches forward and the screeching sound against the ship stops, but *Wild Hair* doesn't break free. We have to wait for more tide, more lift. I am desperate to be free, feeling as patient as a dropping paratrooper waiting for her chute to open.

Thirty minutes pass and the rupturing noise starts anew. I scramble back to my place, rev the engine, and *Wild Hair* strains forward. Dave is again at the bow, on the balls of his toes and shifting his weight to stay loose, a prize fighter before the match. I can tell by the way he's looking and not looking that his imagination is taken by a vision of *Wild Hair*'s keel stuck into the sea floor.

"C'mon honey. Push!" I say to the boat. The prop spins, kicking up sand. The worst grinding noise yet shutters from stem to stern. I gulp. My ship hops, clunks to a stop, shifts, and finally springs free.

"Heather, that one," Dave shouts and points—panic is in his eyes. I look at the winch. The line he draped in big loops around the drum is paying out fast. There's an anchor on the end of this line and we're about to tangle the rope into a mess on the winch. Snagging the works would halt our escape. I lunge. My fingers flip the snarl onto the deck before the mess goes taught, and I watch the line zing over the side. Just like that, we abandon the first anchor.

Bit by bit, *Wild Hair* bucks and claws its way from land. But the bow anchor, which we must have dragged into the shallows as we went aground, now lies to our portside. Driving past it, this second line becomes taut. The anchor at the other end has a tough grip, and all at once its pull makes *Wild Hair's* path arc. In an explosive move, Dave dives to his knees and saws the line viciously with the serrated pocket-knife attached to his life vest. He's a wolf chewing off a leg in a trap. It tears clean through but I realize we've lost a limb—the jagged end (and the anchor it held on the opposite end) disappears in the next wave. Dave kneels and looks at me with tired eyes.

"I didn't have a float on that line," he says.

Damn.

My hands spin the wheel right and left to test the rudder's soundness. The boat turns well, so it appears the rocks didn't harm *Wild Hair's* steering. That's a relief. Also, I don't feel any shimmy underfoot—a good indication that the prop shaft isn't bent. Unexpectedly, a ray of daylight pierces my attention and I turn. At some point, when my focus was elsewhere, morning arrived and the sunrise is glorious: pink, green, and gold. The spectacle is so congruous to the moment, I want to look around to find the audience to this comedy. I half expect a choir to sing an extended major chord to harmonize with the light and a marching band to float past—horns tooting—to celebrate our success. I laugh at the synchronicity of new beginnings—the morning's and mine.

"The storm is over," I say as Dave nears.

"It's a beautiful day, baby! I'm checking for leaks." He disappears below. I grin to the horizon, thinking: we are free and we are safe. I am exhausted.

Out a ways, our third anchor splashes, but neither of us trust the ground to hold the ship. We nap in shifts. Eventually the breeze calms to ten knots and the sea lies down.

Late in the afternoon, Aggghan and his wife dinghy over to see how we made out. Mrs. Aggghan speaks English.

"Alain was wondering if you intentionally reanchored before the storm," she says.

"Wait, your husband's name is Alain?"

"Yes," she says, "Alain."

"Aggghan," Alain says nodding, smiling.

"Oh," I say, feeling dense. "No, we would NEVER reanchor before a storm. We thought we had a good set from the first blow. We wouldn't mess with something that was working."

"He wondered because he saw you late in the day in a new position, closer in."

"We must have dragged before the sun went down and didn't realize—"

"We're going to have to keep better tabs on our position at anchor," Dave says.

"And he wants you to know that he couldn't see you were in trouble. It was good that you shined your light on the island. He didn't know you were grounded until you did."

"Ah . . . I figured."

"It was a good thing you flashed your light when you did, because we were about to go to sleep."

"To sleep," Alain nodded. "Then we see you no more! We don't come."

"Wow," says Dave. "We are so lucky for your help! Thank you, again."

"Thank you for risking the dinghy ride in the storm."

Alain gives a small shrug. He thinks his actions unheroic, simple. But his working outboard saved our boat and maybe even Dave's life. With his wife translating, I tell Alain the full story of the night, including the Mayday. My words peter.

"I'm so glad you were here." My chin puckers and I look away. We sit in silence—four sailors with a common understanding of how bad things can turn in an instant.

The wind disappears and peace settles as the sun sets. The plot of *Lost* lacks thrill for us now. We practice a new habit: each time we stand, we check our ship's location through port holes. At bedtime, the mattress is horizontal and we sleep together as sound as puppies.

The next morning, I locate Charles's phone number in small print deep in the pages of a Bahamian guide book, and telephone on our Bahama cell phone. He is the local man-on-call for BASRA. He arrives in minutes from Little Harbour Cay to help fetch our lost anchors. Charles and Dave zip about with the man's mighty outboard.

"You're never going to believe it," Dave says on their return. "I thought we lost the CQR for sure since there was no float on the line. But we lifted the line for the plow anchor and the CQR's line was draped on top."

"You recovered a thousand dollars' worth of equipment," I cheer. "You—Charles—have superpowers! Party on *Wild Hair*!" The men laugh. "Charles, I know it's only nine in the morning and it's the middle of the week, but would you like one of my husband's beers?"

His eyes look toward the horizon with the seriousness of a sage. "I never did say no to refreshment."

Sipping my morning coffee, I tell Charles our story.

"Dis here is a marl bottom," he says. "You tink it's deep? De sand is nothing but rock with a bit of dust on top. It tricks everybody and their anchors don't hold."

"Charles said he helped salvage an 85-foot sail yacht last month. The professional crew blew aground like us, but they did it at high tide, so they couldn't drive off when the water came up."

After lunch *Wild Hair* motor-sails south. The Northwest Providence Channel is benign—a coiled snake. Deep blue water sloshes in four-foot waves. Droplets sparkle. We arrive at Frazer's Hog Cay in late afternoon. It is a small settlement with a lively marina. Mooring balls dot the harbor. I snag the loose tail swinging from a fat, white float with a boat pole, thread a starboard bow line through the loop, and cleat it off. Before *Wild Hair* swings in the breeze, I thread a second line from the port side. Sailors in dinghies speed our way.

"It's *Wild Hair!*" someone shouts.

"Oh, it's *Wild Hair*," a woman on the next boat cheers. "Look!" she says to the crew of a trawler as she points.

Behind, folks shout, "Hooray!"

A fellow on shore jumps into his dinghy and joins the commotion. About a dozen dinghies collide near our hull like bumper cars.

"Are you OK?" a woman dinghy driver asks.

"Do you have everyone onboard?" a grizzled sailor with an uneven-buttoned shirt questions. A cigarette hangs limp from his lips.

"Is the whole crew accounted for?" a young man asks. His pretty wife looks on.

People keep arriving. My mind spins to catch up.

"Oh my, we were just sick listening to you! We've talked of nothing else."

"Did you make the Mayday?"

"What happened?"

"So, you heard me?" I ask. My face flushes.

"Yes," a half-dozen answer in unison. "Of course we did." The chatter begins again with the enthusiasm of a soldier's homecoming.

"But you couldn't help because it was too dangerous?" I interrupt.

"Of course not!"

"It was horrible out there—"

"That's right. What a night that was, eh?"

"Is that our captain?" someone asks Dave. He looks as overwhelmed as I am. "How did you make it back to the ship?"

"Hey, you made it after all! Boy, are we glad you're here!"

"Yeah," says Dave laughing. "I'm glad to be here, too."

"Tay went out looking," says a Bahamian voice. It is the marina worker responsible for the mooring ball.

"What's that?" I ask.

"Tay sent a boat into the night, looking. But there was just a bit of location. Tay didn't know where to look."

"But I gave our latitude and longitude, many times."

"Tay no hear it all. Tay go out here to look." He gestures toward the Northwest Providence Channel nearby.

"You searched for Dave twenty miles away from where we were?" The floor drops from my stomach.

"Not me," the man clarifies. "But Tay look. Tay try."

My shoulders collapse; my hands cover my face, hiding regret. I am responsible for sending people into that night. I steel myself for more. "My friend, how was the call heard—on the VHF?"

"No, de Coast Guard say, 'go.'"

"The US Coast Guard heard me?"

"Yes." The man is distracted. "You come to de office later 'n pay for de ball." He pushes his dinghy off, waves goodbye, and zips toward the piers." The crowd offers well-wishes and disperses.

I sit at the salon table with a sandwich, soda, and chips, and open my email for the first time in days. The inbox is loaded. Subject headings read: "Are you alive," "Call home," "We're trying not to worry," "Contact your US Coast Guard," and "WTF: WHERE ARE YOU?"

Everyone knows: children, parents, brothers and sisters, strangers, and friends. But what they don't know yet is we're safe. I'm overwhelmed by the thought of all these people worrying about us.

I grab the phone and tell Dave we've got calls to make. Our conversations reveal the potency of my SSB radio calls. Switching frequencies, my voice bounced like a tennis ball to greater and greater distances. For unknown reasons, I wasn't heard in the Bahamas, the East Coast, or on most of the continental United States. Only Petty Officer Adam Harris, manning a US Coast Guard communication station in Kodiak, Alaska, intercepted and recorded my transmission. He picked it up on frequency 4.1250. Static garbled the position. He mapped *Wild Hair*'s location as best he could and mobilized the Coast Guard District Office in Florida. Florida briefed the Royal Bahamas Defense Force and a local BASRA volunteer rescuer—a fellow just like Charles—set off with a friend. Twenty-one miles off course, their effort was futile.

There's more. When the Coast Guard heard my Mayday, they opened a Search and Rescue (SAR) case. Once a case opens, it stays open until the Coast Guard either finds what they are looking for or confirms services aren't needed.

Officer Harris handed our case to a talented detective also stationed in Kodiak: Joshua Bouknight, a petty officer OS2. Bouknight reviewed the national database and discovered twelve vessels registered with the name *Wild Hair*. Then he turned his attention to the recording, put it through cleaning software, and removed the majority of atmospheric noise. He listened to the call several dozen times. Hearing, "One person blown out to sea and one person on board," the officer figured two people were onboard, so he ruled out large, commercial fishing vessels.

Turning to the Internet, Bouknight found my sailing blog. Places I wrote about in our travels matched the vicinity of the Mayday. I also described *Wild Hair*'s length and brand, so Bouknight cross-referenced boat details with the vessel-registration database and deduced our names and contact information.

Bouknight telephoned our old home number, but it was disconnected.

An Internet spider found pages related to my husband and his preretirement job, so Bouknight called Dave's former secretary in Wisconsin. She confirmed we were sailing and gave him our cellphone numbers. We had a travel-hold on our service, so this was another dead end. Dave's secretary knew our children were close friends with my husband's partner's kids, so she put Bouknight in touch with Dave's partner John. John called his daughter, got Maggie's phone number, and gave the contact information to the officer. Bouknight called Maggie. Our daughter must have been terrified.

"Hi, Maggie. It's your mom and dad." Our girl is on speaker phone. "Hi, Noodle," says Dave.

"Well, there you are." Her voice is steady but her tone is uncertain. "What's going on, guys?" We tell her our story and learn hers. She relaxes as we speak. Words begin to flow.

She was in the college library studying for a test with her boyfriend when Bouknight called. "He said he was in Alaska and that he heard a Mayday call that was probably from you, and I thought that couldn't be because you're nowhere near Alaska. That's not how space works. He kept talking and I was trying to listen and be helpful, but I was confused. I didn't know what to do with the information he was giving me."

"I'm so sorry, sweetheart," I say.

"It was scary and I was irritated that the closest person to give a shit about you was in Alaska."

"They mobilized people here too, Noodle," Dave says. "Lots of people tried to help."

"He didn't tell me that. I felt pretty helpless, you know? There was nothing I could do in the middle of Wisconsin. So I told them to talk to grandma because she at least had all your emails."

"You did great, Maggie. That was the perfect thing to do." I imagine her wrestling with fear for more than a day. "Were your friends helpful?"

"No. They were nice and all, kind of supportive, but they really didn't understand. They couldn't relate and say, 'Well, when my parents were out at sea and called a Mayday . . . ' I told them what happened and they said, 'that's really strange . . . so anyway . . . .' It was surreal."

"You were at sea with us," Dave says.

"Yeah." Maggie pauses. "You know, I just never thought about you guys dying before. I mean, I knew you guys were probably going to kill yourselves somehow in your adventures. I just didn't know it would be so soon." We laugh.

"Thanks, Maggie," says Dave.

"But I decided if you died doing something you loved that that would be better than being too afraid to do it at all, you know? I'd be much more upset with you guys if you never went sailing and became lame, old, retired people. I thought, 'At least they died together!'"

"We're not dead yet, ya know. We're going to stick around and make you miserable a little while longer."

"Yeah. Now that I know the big, scary thing happened and you survived, I guess I'm going to worry less. You're not going to let just anything take you."

"Well, that's backwards logic, but I like it!"

Next we call my folks.

"I told the officer, 'This is a mother's nightmare,'" my mom says. "The world stopped; we've just been existing, waiting to hear everything is good. Your call is so welcome."

"Oh, mom. I'm so sorry this happened to you."

"I gave the officer your position and confirmed the *Wild Hair* he was looking for was most likely my daughter. I kept thinking it wasn't you because you wouldn't have frivolously done a Mayday. But it had to be you."

"It was. We were in a hard place," I admit.

"I tried to call Eland, but his cellphone mailbox is full again. He doesn't know."

"Well, there's one person not worried," says Dave.

"Are you going to stop sailing now?"

The question makes me smile—the thought of stopping hadn't crossed my mind.

"No, we'll keep on. What happened is just part of sailing. Bad things can happen anywhere."

"Well, you're adults and we have faith in you."

"And you learned something," my dad adds. "You're smarter, right?"

"Right," Dave and I agree in unison.

My inbox contains at least one email from every person touched by Bouknight's investigation. I draft an email announcing our safety and broadcast our good news with the push of a button and the efficiency of a White House Press Secretary. I also email Officer Bouknight, and we correspond throughout the day.

I tell him I feel shame-faced for wasting three days of his time, and who knows how many other Coast Guard resources. "I didn't realize anyone heard me," I write. "Plus, sailing books talk not about terminating a Mayday but sending them. How does a person stop a SAR case once it gets rolling?"

"The Agency is structured like a wedding cake," Bouknight writes. "It's a series of narrowing tiers. Messages find their way swiftly to the right person so you can notify anyone in the system." He adds, "Mariners should telephone after shooting a flare, sending a voice distress, or firing an EPIRB." I get it. Follow up calls keep rescuers out of harm's way and avoid stretching finite resources so more individuals can receive assistance.

In every exchange, the tone of Bouknight's writing is kind and modest. When I thank him for his perseverance in finding us, he writes: "I was trained to search for every boat like I had my own family aboard. Many people were involved in the SAR, and every one of us is trained that way. Everything was made possible by a chain of individuals who operated with consistency and professionalism."

I share my uncertainty about the wisdom of my decision to issue the Mayday in the first place because the call was based on what turned out to be an optical illusion; Dave wasn't in peril. Bouknight insists my actions were appropriate in the moment.

"All Coast Guardsmen feel only relief when they find a vessel in distress is safe. Too often, our best is not enough, no matter what we do. Too much works against us, be it the ticking clock or the wrath of nature."

He goes on to compliment my carefully articulated message and calm manner during the call (funny, that's not how I remember feeling) and expresses only gladness that we are safe and sound. No one—he says—should hesitate to call for help. I reread his words several times and try to imagine the struggles he's seen, the suffering rescuers carry. His compassion and honesty stun me.

It's then that I notice the Latin phrase included automatically with his signature: "*Dum spiro, spero.*" I do a quick search and find the phrase translates as, "While I breathe, I hope." My body jolts in attention. Who—I wonder—is breathing and hoping: him, me, or both of us?

REFLECTIONS ON POWER

I've had months to reflect now on events surrounding the Mayday, and Joanna Macy keeps popping into my consciousness. She is a fellow spiritual ecologist, Buddhist scholar, and environmental activist who has been around the block about 10,000 times. Macy distinguishes in her writing and teaching between Passive and Active Hope.[29] We are caught in the talons of Passive Hope, she says, when we wish for things to be different but do absolutely nothing to change our circumstance. Alternatively, we engage in Active Hope when we deeply examine our situation, get clear about what we'd prefer to happen in and around us, and take simple concrete steps to bring about a new reality (regardless of our likelihood of success). None of us have power enough to control the forces that be, but each of us has a sovereign power to control what we personally do.

Officer Bouknight didn't sit back and long for things difficult and far away to somehow work out. He and the entire military branch embodied the three fundamental steps Macy advocates. Personnel monitored radio frequencies for disasters and heard my call for help. They were clear about the reality they preferred: not danger but safety at sea. They took immediate and decisive action to rescue Dave. On top of that, they didn't waste time weighing the likelihood of success before they mobilized. Optimism wasn't a factor. They simply followed the recipe for Active Hope—paid close attention, knew what they wanted, and acted as best as they could—even though circumstances looked bleak.

I am so impressed and so very grateful.

In a way, I guess I practiced Active Hope at White Cay too, when I took in the situation as it unfolded, aimed to free Dave from danger, and radioed for help. Of course, I was swamped in a river of emotion, but that didn't matter: I did what needed to be done.

In some ways, I was powerless when the inflatable dinghy blew between the islands with Dave onboard. There was no controlling the storm, the slipped anchor, or the faulty outboard motor. It was impossible for me to reach my man to save him and, at that point, I had no influence over the circumstances leading to the crisis. In other ways, however, I wasn't powerless: I still had dominion over my own actions. My radio call set into motion a rescue mission that didn't stop until I brought the response to an end days later. In the great stream of cause and effect, I had the capacity to mobilize tremendous resources in support of my preferred outcome. Evolution, it seems, has given me and others in my species the rare and wonderful gifts of awareness, volition (or clear aspiration), and choice, and although these qualities don't give us power *over* Nature, others, and circumstance, they do give us the spiritual power to act. Humans can cultivate healing and well-being anytime and anywhere. Almost magically, the powers of awareness, volition, and choice originate from within and don't require anything in particular from outside conditions. Awareness, volition and choice are *superpowers*.

In the midst of my Mayday call, I aspired for something inherently good, a goal that others—like the US Coast Guard—wanted, too. By tapping into the synergy of *collective* aspiration, my influence to make a difference sprouted faster than kelp. The power of my choice also got a boost from a willingness to do something small—radio for help—even when the action appeared in the thick of things to be fruitless. This leads me to believe that even the tiniest deed, set into motion, can spark a revolution.

This back and forth analysis of whether I am powerful or powerless is well worn territory for me. The inquiry was built into my responsibilities as an environmental justice advocate and I know it to be a common preoccupation among most change agents. We strategize endlessly about ways to leverage power over people and situations, because it's our responsibility to stop others from causing social and ecological harm. But in my experience, there is never enough power *over* a situation to overhaul the system perpetuating social and ecological damage, once and for all. And even though advocates lose sleep and our hair researching the nature of our challenge to the nth degree, even though we know beyond a shadow of a doubt our aspirations are altruistic, every power play we make is undermined by an equally forceful maneuver by opponents. Caught in a vast array of power struggles, we wear out our relationships and exhaust ourselves.

It's silly to think we can avoid power skirmishes altogether. Advocates must organize swift and strategic holding actions effective enough to keep things from getting worse. We are the players in society working to exercise control over bad actors by helping, among other things, to make laws that put certain behaviors off limits. But laws are temporary and holding actions ought never to be confused with *transformation* (this is another lesson from Joanna Macy). Somehow, more of us should find a way to direct our personal power toward actions that don't merely constrain harmful actors and actions, but actually transform society's collective consciousness so fewer holding actions are necessary. That is a tall and noble order.

This time, the lesson from the Great Atlantic Teacher seems to be that every action I take has the potential to set into motion a ripple of causes and effects. That's pretty significant. It makes me think through a litany of personal actions I could take, and their possible consequences. Should I buy an expensive suit, or make a donation to fight hunger? Should I eat a hamburger or enjoy a plant-based meal? Should I drive an SUV, a small hybrid, or ride my bike? Should I talk about climate change, or keep silent? How endless are the opportunities to exercise personal power?

There's really nothing too complicated about doing the most helpful thing we can that is within our reach. But simple as it may be, it is very important. I suspect Earth—humanity's witness—takes notice. After all, humankind, armed with modern technology, is the only species powerful enough to shape the planet's fate. It is on us—during this climate emergency—to act as responsibly as we can for as long as necessary until danger subsides. Our optimism or pessimism is immaterial. We can let the stories we tell ourselves go, and exercise our superpowers of awareness, volition and action. We can act with the commitment and clarity of purpose the US Coast Guard brings to saving just one life.

I offer this meditation as a spiritual technology I have found useful in my practice. It helps me cultivate Active Hope and wake up to my innate superpowers.

> Breathing in, I lift into my awareness a condition
> of suffering in the world I cannot abide.
> Breathing out, I look deeply into the root cause of the suffering.
> In, aware of suffering.
> Out, looking at root cause of suffering.
> (Ten breaths)

> Breathing in, I become curious about the circumstances
> feeding the suffering, causing it to persist.
> Breathing out, I see the root cause's food.
> In, curious about circumstances.
> Out, see suffering's nutriments.
> (Ten breaths)

> Breathing in, I aspire for a new reality to come
> into being in the suffering's place.
> Breathing out, I see my volition as something
> that is inherently good.
> In, a preferred reality.

Out, a clear volition.

(Ten breaths)

Breathing in, I see a small action I can take in
    support of an alternative reality.
Breathing out, I accept I have no control over
    the consequences of my actions.
In, see small action.
Out, release attachments to specific outcomes.

(Ten breaths)

Breathing in, I pledge to exercise my superpower
    of choice and take the action before me.
Breathing out, I let go of controlling the outcomes of my actions.
In, exercise superpower of choice.
Out, Let go of outcomes.

(Ten breaths)

PART II
THE PRACTICE OF KNOWING

JACKSONVILLE

JACKSONVILLE, FL TO
CHARLOTTE AMALIE, USVI

THE BAHAMAS

MIAMI

TURKS AND CAICOS

CUBA

FOWL CAY

DOMINICAN
REPUBLIC

JAMAICA

HAITI

PUERTO
RICO

USVI

8 IMPERMANENCE

Once started, the route from Jacksonville, Florida, to St. Thomas, US Virgin Islands doesn't have an easy out. I can't pull over and buy things I forgot to pack, or call it quits and get a hotel room when driving gets hard. Pretty quickly, *Wild Hair* will be past the reach of rescuers and traveling beyond the timeframe of accurate weather predictions. The boat is about to become an isolated speck in the vast unknown. It'll take a while, but I want to sail the fifteen hundred miles nonstop to our destination. Dave and I spent years preparing for this, and the journey will change everything. We will become true offshore, blue water sailors. This rite of passage will mean we have crossed the threshold between learning and mastery.

"There are butterflies in my tummy," I say, unzipping the mainsail's canvas cover. "Are you feeling good?"

"Oh, no," Dave confesses. "I'm fluttery, too." Dinghy—clipped to the helm—keeps her focus on a pair of seagulls that swoop.

This departure routine feels especially rusty because *Wild Hair* spent the past five months of hurricane season hauled out high and dry on stands in the boatyard. We were at our house in Wisconsin. Pulling off the financial stunt of a multi-year sailing sabbatical meant buying down to a small fixer-upper. The house we bought was a wreck. But it was set in a lovely neighborhood beneath colossal oak trees, and the real, hand-cut, stacked-stone, floor-to-ceiling fireplace was a skilled mason's tour de force.

Except for the roof, Dave and I are doing the work on the house ourselves. This means that when we weren't fixing busted things seven months of the year on the boat, we are tearing apart walls and putting up two-by-fours. In my opinion, boat upkeep is a hell of a lot more fun.

"Do we even remember how to sail?" I ask.

"I dunno. Scary, isn't it?"

It's my turn to drive us out of the slip. Dave is in charge of the sail. Mentally, I rehearse a smooth, straight-back reverse from the slip. The turning area is tight and vessels are close. But there should be enough room to maneuver and plenty of depth beneath the keel. After all, we got into the slip, didn't we? When the bow clears the dock, I'll power into a portside turn.

The air is 72 degrees. The sky is cloudless. Somewhere nearby, a manatee spits bubbles through its nose and inhales.

"Are we set?" I ask.

"We're as ready as we'll ever be." These are not glib words. The boat is perfect. We have all the right equipment onboard and most of it is new. The two of us are practiced and well-read. We're feeling healthy and strong.

Dave releases the dock lines. But before I get enough speed for boat control, the river current shoves *Wild Hair* sideways and she twists in reverse into the back of the neighbor's boat, *Serendipity*. Steel davits—beefy arms that extend behind to lift *Serendipity*'s dinghy from the water—skewer *Wild Hair*. Dave fends the other boat off, but lifelines and stanchions tangle. We'll have to phone the office and report the accident—offer to pay if we damaged their boat. I gain steerage and drive off, kicking myself that I wasn't more aggressive with the engine in the invisible current.

I scan the river for traffic, but there is none. I was prepared for anything, and misfortune struck before we left the boatyard. If I were a superstitious type (which all sailors are), I'd worry the entire trip was fated to go poorly.

Dinghy is demanding we return. Dave fetches her blanket, picks her up, and bathes her in soft words.

We're starting twenty-two miles up the St. John's River. Today's destination is a pull-over spot near where the river meets the ocean. There, we will anchor and wait until Chris Parker—our land-based sailing partner at Marine Weather Center—tells us it's time to leave.

~~~~

A twenty-two-knot breeze knocks the peaks off two-foot waves when it's my turn to take the helm again. The falling tide conspires with the river's flow, and *Wild Hair* positively dances with activity as we surf at nine-point-one knots. Leaves hang red and gold along the river, but November's usual biting gusts and snows are missing. I love being at sea, but it feels weird that Thanksgiving is in four days and we won't be with family. The time to sail south is now, between the end of hurricane season and the beginning of the Christmas winds. A choice had to be made between sailing offshore to the Caribbean islands, or spending Thanksgiving together. This is a once in a lifetime opportunity. So, the kids are with their grandparents and I'm cut out of deciding who will bake pies, chop gizzards, and make gravy. I console myself with the words of the diesel mechanic in Georgia, who said: "Heather, you can have everything in this life; you just can't have it all at the same time." Now ain't that right.

Two tug boats are doing figure-eights in the center of the river ahead. What the heck? I switch from autopilot to hand steering and stand up, ready for anything. I see for the first time the immense container ship that had been too big to notice. I've got to beat it. Our anchorage is a hundred yards off, up the creek on my port rail, so I accelerate and spin the wheel sharply and the monster ship narrowly floats by.

Good. We've made it this far. The anchorage is our staging point. The plan is to stay put here until Chris says we have a safe weather window. He's thinking that won't be until next month. We'll be ready. The next land we step on (if we ever step on land again!) will be Caribbean.

Mayport, Florida // November 23, 2010

"I know what I said yesterday, but everything changed," Chris says through static the next morning. "A departure now just might work. With light and variables, you'll motor across the Gulf Stream at zero-nine-zero, arriving to the other side at mid-day tomorrow. Then you'll have winds at zero-two-zero to zero-four-zero at ten to twelve. Close haul it as near as you can." I scribble down his words verbatim.

The passage from Jacksonville to St. Thomas is technically challenging. Jacksonville is nearly the most western spot on the east coast. St. Thomas lies not just 750 nautical miles south but also 800 nautical miles east. The dominant wind is from the northeast and the further south we go, the more the trade winds take over from an easterly direction. This means we have to point straight east at this latitude, while the wind has a northern component. Pointing practically into the wind, *Wild Hair* will buck and fuss, making the trip potentially uncomfortable. In a perfect world, we'll travel as far east as our destination—to 65 degrees west longitude—then hang a right and sail straight south along what cruisers call the I-65 corridor. On this leg, trade winds should simply tip *Wild Hair* at an angle and make her soar smooth, happy, and fast.

Chris is hunting for a calm stretch for us to motor across the Gulf Stream, followed by northerly winds so we can sail east. The problem is, if the wind swings and forces us to angle south too soon, we won't be far enough east to reach our destination. We might be forced to stop west of our destination—possibly near the Dominican Republic. Although the D.R. is a lovely spot, heading east from there is difficult, again because of those pesky trade winds close to the equator. Christopher Columbus tried to sail east along the thorny coast of the D.R. and he wrecked the *Santa Maria*. The *Santa Maria*! I'm afraid of what that coast could do to *Wild Hair* if Columbus wrecked the Holy Mary of Immaculate Conception.

This morning, when Chris Parker says, "This just might work," what he means is we might be able to get far enough east to make it all the way to St. Thomas.

Chris Parker has moved on to coach other mariners. Dave and I stand stunned.

"Oh my God . . . It's time to go," I whisper. Instantly my temples squeeze in pain, so I step into the head to dig migraine pills from the cupboard. I concentrate on slowing my breath.

"He could have said something better than 'it just might work.'" I mutter.

"We can absolutely take the four months you figured for island hopping through the Bahamas, Turks and Caicos, D.R., and Puerto Rico, and into the Caribbean. I don't mind," Dave says. "We can do whatever you want to do."

"Yeah, but four months of dealing with cold fronts versus a quick blast offshore? Let me chart what Chis laid out and see for myself how it looks."

Wild Hair's average speed throughout the trip ought to be about five-and-a-half knots—just over six miles per hour. Knowing speed, the general direction, and the distance we can travel in twenty-four hours, I estimate our location at 07:00 each morning specific to Chris's predicted wind directions. The plan looks doable—like it just might work. I ask Dave to double check all my assumptions and do the math himself. He confirms I got it right. This is our window. It is time to go.

Anti-nausea patches go behind our ears; the ditch bag moves to the salon sofa in case there's a need for a quick escape to the life raft; shrink-wrapped dinners I precooked shift from freezer to fridge; the sail cover unzips; the mainsail lifts; the anchor comes up; and the hull gets busy hunting and pecking its way through the chop. We are underway. The day is sunny and warm and the barometer registers a perfect 1020 millibars. My busy mind frets over 10,000 things that could go wrong, but I breathe and tell myself all is well as the shoreline shrinks to a sliver. The beach has nearly slid over the horizon when my cellphone rings.

"Hi, Mom, it's me," Maggie says. "Are you heading out sometime soon?"

"We're out, way out. We may get cut off and I won't be able to call you back. If we do, don't ever forget that we love you. OK?" I notice my voice sounds urgent and clingy.

"Yeah, OK." Maggie deflects the melodrama and launches into a string of college details. She is upset because a roommate is moving out of the apartment they share and into a relationship with an abusive man; another roommate's cat is using her cat's litter box, so Maggie's cat is pooping in her bed; she is making Christmas presents; it's snowing in Wisconsin . . . Then, silence. We're cut off.

I feel a flash of panic. I inhale a single gulp of air. My shoulders hunch in a primal, visceral response. I look around and take in a blue sky with wisps of clouds. Waves slap softly against the hull. A sea turtle ambles by, making me think *Wild Hair* is in the Gulf Stream. Dave is reading a paper chart at the hull. I stretch and make myself yawn in an attempt to relax and go below to nest the cellphone deep in my sock drawer.

In the late afternoon, I set up the pilot berth. Hands jam sheets around the cushion of the salon sofa—the berth closest to the mast. Like the fulcrum of a teeter-totter, the mast is the most stable part of the boat at sea. In the boatyard, Dave bolted the tail of a cloth under the cushion and I pull it out and lace the opposite edge to handholds overhead. This makes a cocoon for us to sleep in to keep us in place as the boat tosses. Dave and I will hot bunk here—take turns sleeping in this bed, exchanging only our pillows. I slide into the close space, lie back with eyes closed, and try to calm my seasickness. It feels as if I'm lounging on a raft in a pool while a fifth-grade class does cannonballs around me. And this is a calm-ish day.

After dinner, Dave clips Dinghy to her leash and nudges her toward me at the helm. The motion of the boat makes it easy to slip, so she steps with the care of a tightrope walker. Her face is wrinkled like an old man's. Poor thing—she's seasick, too.

"Look at the horizon, kitty." She stares miserably at the floor. I combat a jolt of guilt with reason: cats are sailors. They populate the world today because the Phoenicians put them on ships to eat the rats that would otherwise gnaw ropes and eat grain headed to market. Vikings did the same. I've heard Europeans transported thousands of rat-killing pets to protect the foodstuffs of the colonies. But somehow, I don't think Dinghy cares about that.

A pod of dolphins surfaces near the boat, exhaling noisy breaths through the crowns of their heads. Dinghy leaps to attention. My girl's not too bad off after all.

The sun disappears beyond the horizon behind me. Overhead, the sky is feathery blue and pink. Ahead, dirty lavender is blackening the route. The darkness makes me shudder. I prefer my water brightened by sunbeams. Only a sliver of light skims the surface to illuminate sand-colored swirls. *Wild Hair* is plowing through what looks like glowing frosting. Then, in another moment, light leaves the whirls, and the water disappears altogether in darkness.

30°41'21" N, 78°49'21" W // November 24, 2010
150 nautical miles east of Jacksonville, Florida

I wake to daylight after an incredible five-hour sleep. I feel good. The plan is to rotate watch every three hours. But once a day, we'll let our mate sleep as long as they can to get to REM stage, because traveling days without a deep sleep can destroy the mind.

"I had a horrible scare in the night," Dave says as I climb into the cockpit. "It was awful. A cruise ship came over the horizon toward us, all aglow—you know?"

"Yeah?"

"And then it burst into flames. Exploded!"

"What?"

"It was horrifying. People were burning to death. I didn't know what to do. I almost woke you to send a distress call. It just kept getting brighter, kept burning." I look at him expecting more. "Then the flaming cruise ship turned into the rising moon. The whole thing was an illusion and the joke was on me, I guess."

"Well that's good," I laugh.

"You want to know the really exciting thing? I had a full moon the whole night. We are going to have a nice bright trip."

"That's really good. The dark is too spooky."

"It was beautiful. The conditions were perfect and we made great progress."

"Great. And I'm glad a cruise ship didn't explode. Now, go get your long sleep. I'm in charge."

I plot our 07:00 location: we're right where we should be, so there's no need to adjust course. I write the time and ship's location in the log book. Whoever is on watch will do this on the hour every hour of the trip so that if the electronics up and quit, we'll have an accurate starting point for dead reckoning. For fun, I check where we're pointed; staying on this heading takes us to Agadir, Morocco.

It's time for Chris Parker's broadcast. With the boat on autopilot, I go below. Today's reception is poor, so I call him on the satellite phone and get the good news: he thinks we'll have continuing fair weather. Excellent. I pop my head above to check the horizon, and go below again to turn on the computer. I want to download GRIB files—illustrations of current weather patterns from NOAA's website. Unfortunately, the computer picks now to ignore the satellite phone as its modem. Dinghy jumps onto the table where I'm working.

"Well isn't this a fine how-to-do, Miss Kitty. Hey, can you make the computer work?" Dinghy hunches on all fours low to the table, elbows up. Her butt is facing me. She's not amused. I'm trying to stay cool about the modem. After all, it worked perfectly yesterday. But no modem means no family emails and people will start to think we're dead. And then there's that sound: the bilge pump keeps cycling on about every twenty minutes or so. Water is coming into the boat from somewhere. At this rate, *Wild Hair* will be in pieces by the end of the day.

After several dozen attempts, the computer finally recognizes the modem and I download the files. Dinghy is gone. So, before returning to the helm, I look for her and find her sleeping in a small cupboard in the aft cabin. Her litter is in a box nearby and there's a small pile of puke on top. I reach in the cupboard to stroke her back and she purrs.

"You are the best kitty in the world. You even get sick in your box, you sweet thing. You'll feel better soon."

~~~~~

The promised winds show up on the far side of the Gulf Stream. Once Dave comes on deck, we raise the mainsail. (Our rigging is such that one person has to steer while the other raises the sail on deck. Also, it's smart to have two of us paying attention during shenanigan-times just in case one of us falls in and the other needs to do a rescue.). With the sail up, I can finally shut down the droning engine. *Wild Hair* gains speed. There's only the rhythmic whisper of waves pulsing along the hull. At last, we are a sailboat.

"We just used a good percentage of our total fuel motoring across the Gulf Stream," Dave says. "That's not a problem because we knew we had to. But our tanks are down and need to be full again, so let's dump a few jerry cans of diesel in so we keep the boat ready for anything."

"Now?" I ask, ready for my nap.

"Yeah, now. The sea is pretty quiet. It's not going to stay this way."

I grab an oil absorbent pad and the key for unscrewing and tightening the deck fill while Dave readies jerry cans and the funnel. Back above decks, I clip my harness and crawl-walk into position while Dave organizes the pour spouts. I sit to pry open the deck fill cap and then guide the pour spout into the hole when Dave lifts the heavy six-gallon jug.

Diesel fumes fill our nostrils. The boat heaves. I begin to sweat.

Nausea clenches my throat.

"I was feeling good, but now I'm going to be sick," I say.

"Keep it together, sweetie. I'll pour as fast as I can." The giant container gurgles bit by bit. I continue kneeling on the bouncing deck, stealing an occasional glance at the horizon as I watch for spills.

Dave struggles to keep his balance through an irregular bounce. We repeat the routine with the second jug, clean up, and relash equipment. Exhausted and ready to vomit, I go below.

<center>~~~~~</center>

After dark, we change shifts. Dave lifts Dinghy from his lap and places her onto mine. Then, he peels off a status report on our progress. "We are trimmed well for this tack and the wind should stay constant. There is nothing out here. We've been averaging about six to six-and-a-half knots at a one-zero-four heading. Every once in a while I can point the bow up to zero-nine-four, but then we lose speed. So I don't recommend it. We're moving well."

"Got it. I'm looking forward to the moon."

"Yeah, it'll be here soon. And guess what? I got hit by a flying fish. Or, actually, I would have been hit by a flying fish if it didn't first smack into the glass of the cockpit enclosure!"

"Holy-sha-moly."

"Yeah, I was minding my own business, when WHAM. Dinghy was fantastic. She spotted it just before contact and leapt to catch it. She was like a Secret Service Agent taking the bullet!"

"Good kitty."

29°57'18"N, 77°30'20"W // November 25, 2010
360 nautical miles east of Cape Canaveral, FL

That constant wind Dave described has left, and I'm pissed. The breeze swung to our nose, so we spent the past twelve hours tacking from southeast, to northeast, to southeast, to northeast. Trying to coax the boat east, not too far south, and not fall short of our destination and suffer the fate of the *Santa Maria* has me stressed. All these hours of intense concentration have led to almost nothing. Charting our location at 07:00 I see we sailed sixty miles in twelve hours but only gained twelve miles of forward progress.

"Twelve miles? *Twelve miles!*" I shout at the wind. "You're killing me out here with your stupid easterlies. Why don't you give me something I can work with?" My patience is gone.

"I slept only an hour last night because of you and the racket Dave made tacking and winching!" I'm panting from yelling. The wind continues doing its thing without showing so much as a glimmer of compassion or remorse.

I'm also upset because Chris Parker delivered troublesome news: tomorrow night, we will get clobbered by a cold front. A gale will overtake us with at least forty-knot winds. It'll be a fierce blow, too strong to stand up in, the nastiest weather we've ever sailed in. The good news is it should strike from behind, so *Wild Hair* ought to be able to surf the waves. The bad news is we might get going too fast, and some of our equipment could tear and break from strain. We need a way to slow the boat down if need be. So today Dave and I will devise a system of chains to trail from the stern to create drag. Whatever we come up with, it'll have to stay put and not go off accidently, but be easy to launch in dangerous conditions.

The worst thing of all is it's Thanksgiving. I miss my family. Right now everyone is probably padding about in their pajamas at my brother's house, looking for coffee cups, laughing with morning breath. Eland is no doubt slicing my mother's cinnamon-sour cream coffee cake. Marching bands and inflated superheroes are rolling across the muted TV. My brother, trying to provoke a response from Maggie, is dangling turkey innards above her coffee mug. And here I sit, exhausted and frustrated and yelling at blustery weather like someone who forgot to take their antipsychotic medicine.

I find small comfort in remembering my plan to evoke the spirit of Thanksgiving on board. It involves canned turkey, canned green bean, and a box of stuffing made into a casserole. I've got cranberry sauce, powdered mashed potatoes, and store-bought turkey gravy, plus two slices of pecan pie currently thawing. No doubt it'll be a poor substitute for the usual feast, but it'll be right for our circumstance.

Two hours passed since my last tack, so I prepare to come about again. I glance at the gauges to calculate the next angle, but they tell me things are different. The world around me shifted and now it's possible for *Wild Hair* to sail at a clip on course. The apparent breeze is a moderate twelve knots from the southeast at one-five-zero. The transformation was silent and sudden.

"Thank you," I say to the breeze. "Now we can move. Was that so hard?" I bump our course to zero-nine-zero, trim the sails, and *Wild Hair* runs like a kid chasing seagulls on the shore.

"I'm going to troubleshoot the leak before the storm comes," Dave says, delaying the start of his turn at the helm. "You OK?"

"Yeah, I'm good. The leak is getting worse. The pump's running every five minutes and it takes to the count of eighty to shut off." Dave's head disappears when he kneels. Shortly, it pops up again.

"You're not going to believe it. It's so cool!"

"What?"

"You know the secondary bilge pump, the one that kicks on if the water rises higher than the first pump?" he asks.

"Yeah—"

"The through-hull's one-way valve is stuck open. And because the boat is heeling to one side, the through-hull is actually below the water line and syphoning the ocean in."

"Aren't you the smart detective. That is cool."

"Why don't you tack the boat and lay her on her other hip, so the valve is above the waterline, and I can make the repair?"

"Roger that," I say, tacking through the wind.

Within moments, Dave stands again.

"Done!" he says. I bring the boat back on course. "The good news is we are no longer taking on water. The bad news is the valve couldn't be saved. We're going into the storm down one bilge pump."

"OK. That means we still have two electric and two manual pumps in play."

"Yup. Ya gotta love redundancy," Dave beams.

With *Wild Hair* on autopilot, Dave and I make the boat stormproof. I secure the coffee pot and put sharp knives and silverware into drawers. I barrel-bolt the freezer and refrigerator lids shut and screw floorboards in place so their contents—if the boat should roll—don't kill us. Dave double-straps the dinghy to the deck and adds more ties to diesel and fresh water jerry cans. Then we unearth buckets of chain from the lazarette, the storage compartment in the transom, to make a towing warp system and lash them at the stern; they are ready to deploy in a way that shouldn't chafe the fiberglass if we need to slow the boat.

As I go below to sleep, Dave says, "I've left a book on the bunk. Would you reread the chapter on heavy weather sailing so both of us have options fresh in our heads when things get nasty?"

"Yup. Good idea."

I have other plans, too. Moving handhold to handhold as the ship bucks, I stuff a shampoo bottle, conditioner, washcloth, soap, and a comb into the waist of my pants. It's day four and past time for a bit of personal hygiene. My flesh giggles at the touch of soap.

~

The Thanksgiving dinner I make is beyond bad; it's inedible. I tip our plates into the sea and unearth peanut butter and jelly for dinner.

At 18:00, we call my brother's home just as the family is sitting down to dinner. On their end, they pass the phone person to person around the table, giving us a chance to exchange words with everyone. I can't relax into the conversation because I'm so nervous about conserving our satellite minutes for weather downloads and a possible emergency. But touching base is enough for us to tell each person individually how much we love them. When I finally click off the satellite phone, I'm left with the hollow of absence. *Wild Hair* bounds in big leaps.

"This is the hardest day of the year to be apart," Dave says. I look at him. His eyes are small and sad.

30°22'21"N, 75°45'28"W // November 26, 2010
240 nautical miles north of Abaco, Bahamas

The salon spins in a psychedelic circle. I've never been too dizzy to stand before this moment. Falling, I reach for the pilot berth, wiggle my way inside, and close my eyes.

I'm on the birth control pill. In our round-the-clock schedule, I forgot two days ago to take one, so yesterday—as one does when you miss a day—I took one in the morning and one at night. Dizziness and nausea are side effects, and I just spent the past six hours throwing up. I'm *way* too dizzy to take my watch. Poor Dave. He must be exhausted. I drift back to sleep.

It's noon. I'm dog-tired and dehydrated, but I'm able to stand. The sea rolls in tightly timed twelve-foot mounds. Our motion is similar to jumping on a trampoline with another person: there is a regular bouncing pattern, but occasionally a super bounce shoots us at a great height in an unpredictable direction. One-handed, I put freeze-dried chicken noodle soup into two mugs and fill them with hot water from our press pot. When I grab for a spoon, a wave sets both mugs flying. Yellow slime drips from windows, woodwork, seat cushions, and between floorboards. Noodles drape like Christmas tinsel from the companionway steps. I whimper in frustration.

"What's wrong?" Dave calls.

"Two mugs of chicken soup just bit the dust."

"Oh—quite the mess. My cottage cheese went flying this morning. It was like something in a B-grade horror film."

My man is wounded as well. Early this morning, he sat in the cockpit strapping on his lifejacket and tether when a super bounce shot him halfway down the companionway steps. He was lucky he didn't fall the distance and break his neck or slam his temple on the way past the corner of the passageway. He didn't kill himself, but he did put a deep gash over his left eye that refuses to stop bleeding. If we were back home, I'd have driven him to the hospital. But we're a long way from home, so I tore through cupboards on the hunt for a suture kit and tried to steel myself for putting needle and thread through my husband's flesh. We didn't get that far, though: my surgeon-husband forgot to put a suture kit in the first aid bag. Oops. So, I went with plan B and made a butterfly from medical tape that I strapped over the slice, after smothering it first with antibiotic cream. Now, Dave's wearing a green knitted cap to put pressure on the wad of gauze over the wound.

The chicken soup was medicine for us both.

I'm thinking we should have done what other cruising couples do: take on crew when sailing offshore. This twelve hours on, twelve hours off routine is tough stuff. If there were three people on board, we'd helm no more than eight hours a day and get more rest. And if someone had a problem and wasn't able to take their shift, the full responsibility of managing the boat wouldn't fall on the last man standing. I'm convinced. Next time, I want a third.

The air is blowing strong—a constant twenty-two knots gusting to thirty. The water moves so aggressively against the hull that it feels as if an entire men's basketball team is bouncing on my trampoline. I sit near the wheel and fight to stay upright. Dave is trying to sleep, but—stressed by the responsibility of it all—he tells me he hasn't slept once since we left shore four days ago. He seems befuddled to me. And now, as he's trying to get some sleep, the contents of every floorboard, cupboard, and nook and cranny fling from port to starboard in the waves. Usually I can adjust course to smooth out the ride. This is not one of those times.

The sun sets and the storm nears.

30°22'55"N, 73°41'00"W // November 27, 2010
325 nautical miles northeast of Abaco, Bahamas

It is 05:15 and—according to the radar—the storm cell is three miles away and closing fast. It glows yellow and red. It's gigantic. Dave is finally asleep and Dinghy is at his feet.

"Now, honey, you pick now to snooze?" Only the compass hears my question.

The first raindrops start to hit hard, like clay flung from a potter's hands. Almost immediately, a blast knocks *Wild Hair* over thirty degrees. The boat stays tipped as the air accelerates more and the canvas enclosure around the cockpit begins to shutter wildly. The rattling is terrifying. Instruments indicate sustained winds are at thirty-five knots and the gusts are blowing to forty-five. It's a strong gale by the Beaufort Wind Scale standard.

I keep reading the gauges to calculate if the ship is in danger. We are moving *over ground* at twelve knots according to the GPS. That's definitely on the edge of too fast for this boat. But, looking at the second speed indicator, I see we are traveling at just seven knots *through water*. The storm is pushing the water and boat together, so *Wild Hair* is gently surfing the waves rolling in from behind. This means our ship is not careening beyond the zone of safety but cruising along at its happy speed as if it were an ordinary, sunny day. Still, the roar of wind and rain make it sound as if *Wild Hair* is shredding in helicopter blades. I take a breath, keep watching the gauges and decide there's no need to deploy the chain warp. Dave can stay asleep.

Something really weird is happening, though. The feel of the boat is off. The pull of the rudder, the way the boat wants to pivot on the mast and turn up into the wind tells me there is too much mainsail flying. I have to keep hand steering away from the wind. If for some reason *Wild Hair* should end up parallel to the waves, things could get dangerous fast. Waves striking *Wild Hair* on the side could swamp and spin the boat like a rolling pin. We would flip over, breaking the mast, tossing everything upside down, and drowning me in the cockpit. I need to resist the pull and crab *Wild Hair* firmly downwind.

The rudder action makes no sense because the mainsail is gone. Dave and I dropped it before the storm. It's lashed to the boom. I do a quick double check and confirm the sail is still secure. We also rolled in the forward sail and left only a miniscule, three-foot-triangle of headsail exposed. The only thing I can think of to explain the pull of the rudder is that the cockpit enclosure is acting like a sail. Its windage is making the boat turn, forcing me to compensate by turning the wheel the other direction.

The edge of the front passes in about fifteen minutes. Everything tied down stays put. The front fades into a gloomy morning and the wind settles down to ten knots from three-four-zero degrees. Dave and Dinghy continue to snooze.

At 06:30, I hail Chris on the SSB to ask if we've seen the last of the storm.

"Let's see," Chris says. I imagine him in his home, controlling a myriad of computer screens and weather models. He's the wizard behind the curtain.

"Uh, no. Unfortunately a second front will arrive at your location at about 12:10. Then you will be out of the woods." His voice is compassionate and encouraging. "You'll have to wait until the second front blows past. The good news is that it won't be as strong as the first."

"Roger that, thanks Chris."

〜〜〜

It's noon and I am at the helm, again. Dave wishes me luck and heads back to bed.

"Wait a minute. How can you sleep? These fronts are big deals. Don't you want to be here to witness the action? It's amazing."

"No. I'm too tired and the boat is set. Besides, you have time-in-type, as we say in the flying world. You are the pilot who knows how to handle the vessel in a storm. Why would I step in now?"

"Alright. But you're going to miss the fun."

"I think I'll call you Stormy from now because you're the master of storms. Goodnight, Stormy."

The second front arrives at 12:35. I watch, nonplussed, as the wind climbs to a mere twenty-eight knots with a bit of rain. After a few minutes, a gentler breeze presses from the north, and the sea lies down. I unfurl *Wild Hair*'s forward sail and point us east again.

30°12′56″N, 71°36′16″W // November 28, 2010
358 nautical miles northeast of Abaco, Bahamas

"I think it's Thursday," Dave says, "because we left the boatyard on Sunday, right?"

"No, it's later than that. It's Friday, for sure."

We volley the point back and forth. Dinghy watches us, but keeps quiet on the subject. Dave starts the computer to download GRIB files. It's Sunday.

Day six and we are *finally* making magnificent progress. A perfect wind is blowing twelve to fifteen knots from the north-northeast—zero-three-zero—and *Wild Hair* is cruising smooth and fast and as quiet as a monorail. She's tilted to starboard and all the cupboard contents are staying put. This was how I imagined the entire trip would be.

Finally, Dave is caught up on sleep and energized. He's tossing garbage, making food, rearranging the sail plan for tonight's forecast, and charting our progress. I watch him while lying on my side in the pilot berth. Dinghy pounces on my feet so I grab her toy, wiggling the wire to make the mouse come alive at the other end. She fills with glee.

"Wow. We are six nautical miles off from where Chris said we'd be way back on that first day. Damn he's good."

"So we didn't lose that much ground on our day of tacking. That's encouraging."

"Would you like to see my broken toe?" Dave says from beneath his woolen cap; the bloody bandage above his eye flashes white and red.

"Why, certainly," I laugh. "I'd love to see your broken toe." He explains that yesterday, while climbing the steps, the boat took a super bounce and he cracked his big toe. The digit is black and nasty. "I have never seen a toe look worse."

"It looks horrible, doesn't it?" Dave gleams with pride. He is crazy adorable.

Today, hundreds of miles from any shore, in fantastic conditions, we sail by listening. If conditions should change, the pitch and motion of the boat will change. We're so in tune with the sound and feel of the vessel that any disruption will be obvious. But for now, *Wild Hair* is happy. We are happy. So we eat lunch together at the table below decks. We read, cook, and write. It feels good to relax our guard and let *Wild Hair* do what she does best.

After a while, I decide it's time to refill the press pot with boiling water. This is something of a Cecil B. DeMille production. First, I set the open press pot into the galley sink and the full tea kettle on the stove to boil. Then I unearth my waterproof bib overall pants, rubber rain boots, and foul weather jacket. I have to roll around on top of the bed in the aft stateroom to pull on my pants, because I can't manage to get my feet into the pantlegs while simultaneously holding on to keep from falling over. Dressed in foul weather gear, with Dave and Dinghy siting well clear of the action, I wedge myself against the counter, slide on the oven mitts, and pick up the boiling tea kettle. Concentrating my attention, I pour the water into the thermal container. Water hits the rim and shoots across the counter.

"Now, watch out there," my grandpa chides, his voice loud and flawless. I startle. How could this be? The man passed away a decade ago, but I hear his voice as clear as if he were standing next to me. Dave is busy in the forward stateroom.

"I'm trying, Grandpa. Don't you see my coveralls? It's going to spill and I know it."

"Not if you're careful." My grandfather always had words of caution. That was his thing.

"I'll try." I concentrate and take my time, but again hot water skirts the edge and deflects away, this time into the sink. I keep at it, but I only manage to fill the pot half full.

"Grandpa, your submarines went beneath the surface," I say out loud. "You didn't have to deal with these waves all the time." He's quiet. I suppose he's thinking of an answer. "I'm sorry, I did my best." He doesn't say a word.

My clothes are loose. I must be losing weight with all this throwing up and lost appetite. It's a diet I can't recommend. I make a cup of hot chocolate. As I wait for the mug to cool, I notice my arm is gimbaling with the boat's motion. Inside I sing, "Hey, Ho, Way to Go; A Pirate's Life for Me!" Then I think, of course. Movie pirates aren't conducting the orchestra with their rum. They're swinging their arms to keep the contents at sea from spilling.

"The fresh water pump just broke," Dave announces. The fresh water pump pushes drinking water from our tanks through our faucets. "Use the manual foot pump to access the tank water until I can get it fixed, would ya? If that doesn't work, we can crack into the gallons of bottled water under the forward berth."

"Or we could use the jerry cans lashed on deck," I add. "We won't go thirsty."

"Not yet, anyway."

29°04'54" N, 69°19'31" W // November 29, 2010
460 nautical miles northeast of Abaco, Bahamas

In the middle of the night, the seventeen-knot winds from the north swelled to twenty-four knots from the east-northeast, making *Wild Hair* rip on the edge of control. Chris didn't warn us of this change, so *Wild Hair* is flying way too much sail. We should reef the canvas smaller, but neither of us wants to go on deck because there's little chance for rescue in the turmoil of rough sea and darkness. We'll have to reef if the weather worsens. But for now, Dave drives headlong into big seas. The boat's motion is a carnival ride gone mad. Waves the height of our house in Wisconsin curl and smash overhead.

Nothing on this voyage is the same one minute to the next. Benign sailing conditions deteriorate. Storms come and go and the perfect sail plan is no longer applicable twenty minutes later. The wind direction clocks in circles around the boat and we're constantly adjusting our direction and the trim of our sails. The boat's motion swings from quiet and steady to earsplitting and feral. Equipment breaks and fuel tanks don't stay full. One minute I feel good; the next, my body is weak with seasickness. Dave's energy vacillates with sleep and his strong frame keeps getting injured. All this chop and change reminds me of the Buddha's teaching on Impermanence: all that arises will cease. My mother said it another way: "All good things must come to an end." This, I guess, is the true nature of reality: change.

"How are things?" I ask, climbing the companionway for my watch.

"We're fine. The boat's fine. I want to keep sailing. Go back to bed." Dave is hand-steering in big arcs to maintain course.

"I should take over so you can rest—"

"No. I'll be happier at the helm keeping an eye on things. So far, we're okay."

"You're assigning me to the hell realm below. It's terrifying down there, you know. It sounds like we're in the depths of disaster."

"I know. Do your best to sleep because I'm going to be tired in the morning."

"We'll change the sail plan in daylight, then?" Another breaker crashes onto the Bimini overhead and gallons of seawater pour through the gaps, dousing notebooks, snacks, binoculars, a flashlight, chart, and layers of clothing.

"Yeah, or I may call you sooner. Go back to bed."

I lie in the pilot berth, but sleep? *That's* not going to happen. Next to my head are the canned goods and they thunk with each toss. Above the cans, dishes clatter and whack. Overhead, hanging ladles, whisks, and serving spoons scrape and knock as they swing. But the sound that gives me indigestion are those of *Wild Hair* herself; she's groaning and popping as if she's coming apart. I know all this hard sailing is twisting and flexing the hull because the seal in the salon window popped, and icy sea water is now dripping through the breach to soak my body and bed. This is a new low. I know we're safe. But I feel worn out over being so damned uncomfortable.

I've lived on *Wild Hair* for three years and I'm starting to understand sailing both is and isn't what I thought it would be. Yes, sailing can be romantic: hoisting sails, dropping anchors, navigating a course, studying the weather, and discovering remarkable destinations. What I didn't expect is that sailing also entails tolerating discomfort, managing fear, worrying about the ship when we're separated by a thousand miles, and learning how to do things like can food, change oil, install copper foil to make the SSB radio work, refinish teak, and coat the hull with toxic anti-foul paint. Who could have guessed I'd spend hours negotiating insurance and dock leases? And I never envisioned that I'd be tossed around like ideas at a strategic planning meeting for days on end, that I'd be exhausted and wet and incapable of doing anything other than wait for things to change. I take a breath and feel my body jostle.

A rose isn't a rose and therefore it is a rose—or so the Buddha said. This is one of the cornerstone teachings that seems somehow relevant to the moment. The way I understand it, the Buddha meant we carry a concept of a rose around in our head, and our idea is based upon a pretty shallow understanding. But a rose is more than a flower with thorns. It contains nonrose elements, like water, cloud, earth, sunshine, gardener and even the gardener's great-great-grandmother. When you really stop and think about it, the rose contains the entire universe—everything exists in a continuum, a great web of what Thay describes as this-is-because-that-is. Nothing, including the rose, can be pried from the rest.

When we understand that, then we deeply understand the thing we call a rose. A rose is not a rose. Therefore, it is a rose. Get it?

I'm on a roll. Sailing is not sailing, therefore it is sailing. My idea of sailing was much smaller than the reality of sailing. Sailing includes everything leading to this point, all the conditions of this moment, and all that will follow from sailing's influence. Everything is related in some way, shape or form, and everything is also temporary, impermanent. It's like the whole cosmos is a massive, undulating, connected goo (that's not too difficult to imagine in this moment). So, if all that we experience is part of one, cohesive, constantly morphing reality, then it is the nature of things to be ultimately unknowable from the limited place where we stand. I'd do well to hold my conceptualization of the world and its component parts loosely, and expect that things will ebb and flow of their own accord. I'd be wrong to think the conditions of this moment (or any moment) will stick around.

I like this line of contemplation about the nature of reality. The overwhelming profundity of it distracts me from the mundane repetitiveness of the clatter. And here's the really neat thing, I think: in this changing stew of impermanence where the most unlikely ingredients are connected, I can see that if one thing changes, the condition that is arising for countless other related things may shift. Taken from a different direction, this means that impermanence and the interrelationship of all things is the root of transformation, creativity and freedom.

That's a big idea to chew on. If my current discomfort is guaranteed to change, I wonder which impermanent element will shift to bring an end to my misery.

"Turn south now," Chris Parker says in his morning broadcast. A-*ha!* There it is. The boat's violent motion will stop because our direction will change. Adjusting course, we will stop pounding headlong into oncoming waves, hammering ourselves silly. We'll run with the wind and sea and slide steady through the troughs.

"You're not at sixty-five yet," Chris continues, "but you'll have another chance to go east in a few days. Stop beating; turn south." I relay the good news to Dave and he spins the wheel. Instantly, the boat tilts and settles. The ride flattens and the cupboards quiet. *Wild Hair* is transformed into a magic carpet.

"You should have turned sooner," Dave's father's voice says with a precocious snicker. Good grief—another dead person, I think. This time it sounds as if Bob is standing behind me, speaking plain as day. He's being his usual self, stating the obvious and finding glee in another's stupidity.

"Oh—hi, Bob," I say aloud. "Yup, you're right, again. Hey, where were you earlier when your suggestion to change course might have been useful?" He says nothing, but I know he's smiling. Dave's dad—a fellow sailor—must be proud of us. Really, I'm glad he's here.

Distance-wise, we're more than halfway to St. Thomas.

My body is falling apart. Medicine in the anti-seasickness patch behind my ear has blurred my vision so I can no longer read the chart. It's a common side effect. On top of that, my voice is gone. Dave thinks I have a fungal infection from using saltwater soaked earplugs. I'll have to learn to sleep without them.

"The boat stinks," I croak as we end a meal. "Let's clean up." With the autopilot on, we wash dishes, bleach the heads, empty the cat box and lasso a bag of trash to the transom. We bathe our bodies, wash our hair, and shave. We stow our soiled clothes. These actions soothe our noses and do wonder for our morale. Then we start above decks, scanning the lengths of our lines for chafe, and refastening jerry cans and other knocked-about equipment. Dave replaces the noisy fan at the pilot berth with a whisper-quiet one; now we can keep cool as we sleep without adding to the din. He also repairs the fresh water pump and water flows from our faucets anew.

As the sun sets, we change the sail plan for the coming night. Winds are accelerating, so we drop the mainsail and leave only a sliver of forward cloth. With hardly any canvas flying, *Wild Hair* is cruising at six-knots-per-hour. That's enough speed for the night.

27°10'13" N, 68°26'40" W // November 30, 2010
475 nautical miles east of Abaco, Bahamas

It is day eight. Five hundred and twenty miles separate us from our destination and I realize (why it took me so long I don't know) that Dave and I enlisted in a sailor's boot camp when we signed up for this trip. I'm numb from the endless raising, lowering, reefing, furling, winching, and letting out. Beyond thinking, my actions are mechanical. Today I sat watch from 01:00 to 03:00 and 06:00 to 10:00. Between shifts I sank into deep sleeps; I woke at 13:00 dreaming I was in a hermetically sealed tube being sucked through a chute at the bank.

Last night, at 03:00, there was mutiny aboard *Wild Hair*. Dave must have slept well because he scurried about the deck changing sails in the dark as I waited zombie-like for my watch to end. After completing essential tasks, he announced he was going to do some nonessential stuff at the bow.

"No. Absolutely not," I said, standing to go below.

"Look, Heather. I'm trying to maintain top performance on this vessel to make the trip as short as possible. A shorter trip is a safer trip."

"No, you look, Dave. I am exhausted. Delirious. Stupid-tired. There is zero chance I could save you if you went swimming. To use your reasoning as a surgeon, am I going to do elective surgery in the middle of the night? No. I'm going to wait and do elective surgery in daylight, after my sleep. Am I clear?"

"Oh," he muttered, looking confused. "All right."

Shit, why don't I ever have the stronger body? Why must I always be the one to broadcast when I've reached my limit and rain on his parade?

~~~~

It's afternoon and time to take another watch. I lie in bed with eyes shut, positively dreading the thought of standing. I've never done anything so physically demanding. Every one of my movements requires concentration, strength, and balance. The most simple tasks are hard work. I don't want to complain, but I've got what feels like shaken baby syndrome. Neck muscles don't want to hold my head upright on jostled shoulders anymore; I'm tired of being a bobble- headed doll. Getting to my clothes six feet away requires my hands grabbing fixed objects, pulling and straining my arms, and my feet stepping in time with the boat's heave. Brushing my teeth at a narrow vanity that rocks wildly requires me to plant both feet on the floor, bend my knees, press my butt to the wall, bend at the waist, wedge my left hand against the basin, and use my right hand for brushing. *Wild Hair* never stops moving, so the struggle to stay upright doesn't stop until I lie down. I am beyond exhaustion and ache all over.

But here's the really funny thing: I am happy. Every time I look, I see a little Buddha inside and she's smiling, delighted we're on this adventure. The sensation is one of being detached from my feelings of discomfort and the separation makes it possible to keep joy flowing. Yes, my body is tired and dirty, my tongue tastes sour with sleep, and there is no doubt my aches will worsen the instant I try to stand. But my spirit is tapping into some sort of transcendent endurance, and I'm able to float above petty circumstance and celebrate this opportunity to sail with the Great Atlantic Teacher.

I'm finding a kind of a refuge in a happiness that exists beyond forms, sounds, smells, taste, touch, and my ideas. The wise woman within seems almost bored by my body's oh-so-predictable but temporary discomforts; it's no longer buying into a "poor me" saga. It seems like I've stopped being pushed around by the tyranny of my senses. Holy cow. Pain is not pain, therefore it is pain—an illusion that blows away like dust as I become comfortable with the formlessness of a reality greater than my perceptions.

Transcendent endurance. I suspect marathon runners experience something similar.

I remember Thich Nhat Hanh saying formlessness is the mother of all Buddhas. Now it's clear that even though pain can feel all-consuming, there's space and freedom beneath pain, a spaciousness where the energy of life emerges unfazed. I am riding uncomfortable waves of continual change and I'm feeling strong, whole, and invulnerable as I go.

~~~~~

Hours pass and I'm at the helm again. The autopilot beeps twice when I adjust course a couple degrees. Thoughts on impermanence and the interrelationship of all things swirl lazily in my head and a thought bubbles: the ocean is not the ocean; therefore, it is the ocean. Now here's something to explore. Sure, the ocean is saltwater, sand, waves, fish and coral reefs, but now I know it's also made of nonocean elements. But what? A vague memory of Environmental Science 101 floats into focus. The ocean is a "heat sink"—*the* reason Earth is not an icy rock hurling through space; saltwater absorbs the majority of the sun's radiation and spreads warmth (as it stirs and evaporates) around the world. The ocean is Earth's thermal regulator, but now that it's overheating, it's regulating less . . . well . . . regularly. Fish are adapting to the changes by migrating toward the poles, and this upends fisheries local economies, and food production. The ocean is a life-support system and it determines if and where life thrives.

The ocean is also weather. Nearly all raindrops falling on land come from the sea. Evaporation leaves salt behind. The ocean waters our crops, makes rivers, and fills municipal drinking water reservoirs. As the climate warms and evaporation intensifies, the sun draws ever-increasing amounts of moisture from the ocean's surface like a giant sucking carpet cleaner, and this pulling saturates the atmosphere. Unfortunately, water happens to be violent storm food. With climate change, the ocean is a manufacturing and distribution plant for cataclysmic storms.

The ocean also creates the currents circulating the planet. Saltwater is set into motion not by a mechanical heart but by surface winds and disparities in water temperatures and saline densities. Powerful conveyor belts of currents, like the Gulf Stream, cool the equator and warm the poles in what's called thermocline circulations. Like specialized cells in a body, life on Earth has adapted to a particular pattern of circulation. But the established flow is changing as the planet warms. Research says a three-plus degree Celsius rise in global temperature may completely disrupt the ocean's circulation system and make temperatures around the world flip-flop wildly between too hot and too cold. If things got this bad, no one would be able to grow crops. The health of our oceans is the health of our agricultural systems.

What's more, the ocean is ice. Some of the ice is in the form of inland glaciers. As this ice melts, the sea level rises. Other ice—like the Artic—floats on the sea, so it doesn't add to rising sea level when it melts. But it is important to keep the floating ice where it is, because that ice contributes to climate stability by reflecting sunlight. When floating ice melts, the dark blue ocean absorbs the sun's warmth. Melting ice turns to water, and the water traps heat and this melts more ice, which feeds a cycle of runaway climate change—the point of no return when processes beyond human control self-accelerate and conditions shift from habitable to uninhabitable for humans. The loss of ocean ice is a negative feedback loop for climate change.

No one is in sight, but still I think of the ocean as transportation and trade. I see in my mind's eye vibrant cities hugging the waterfronts of continents not because the sea is beautiful but because somewhere in the middle of the mix of people and noise is a natural, safe harbor—a place for ships to come and go, where goods can be exchanged and profits made. Over generations, in these centers of commerce, networks of infrastructure evolved. Unlike their inland counterparts, coastal cities appear to me like jellyfish with long tendrils of shipping lanes, train tracks, roadbeds, and plane routes. The spindly hub is kept alive with pipelines, underground cables, and fiber-optic communication networks. Of course, security equipment and personnel—our nation's military—keeps everything safe. Port towns are incredibly complex centers of activity, and rising sea levels threaten to put them under water.

Climate change means ruined infrastructure, lost goods, logistical chaos, and a security nightmare.

Strange as it may sound, the oceans capture and store fifty times more carbon dioxide than Earth's atmosphere. Already, nearly 150 billion tons of carbon have sunk beneath the surface of the sea since the start of the industrial revolution. For decades, the settling of carbon into the ocean dampened the effects of extreme carbon emissions. Unfortunately, the chemical composition of the ocean is just now crossing a critical tipping point. The seas are 30 percent more acidic than normal and dissolved carbon is choking life from coral reefs—the most biologically diverse places on earth. We're conducting a global chemistry experiment that is softening the skeletons and shells of ocean animals. Moreover, scientists discovered that carbon-loaded seawater traps naturally occurring sulfur that would otherwise have become airborne. Earth needs atmospheric sulfur to reflect the sun's rays and keep itself cool. The ocean, a guardian protector of life on Earth, is nearing the end of what it can absorb.

Earth is a blue planet. Her oceans blanket seventy percent of Earth's surface. And it turns out the seven seas provide free, essential, and heretofore underappreciated ecosystem services. The ocean, however, is impermanent, and cannot be taken for granted.

~~~~~

At sunset we are in another predicament. The big storm, with forty-five knot winds that overtook us three days ago, stalled—unawares, we sailed into it again. Chris Parker didn't warn us, so our sail plan is not right for the sudden thirty-seven knot gale and Dave—trying to steer a safe course downwind—is sailing *west*. We're losing hard-fought ground. *Wild Hair* is noisy again—sounding cymbals and drums like a one-man-band. For an instant, I am pissed off at the noise, motion, and forecasting screw up. I start to tell myself a story of anger and blame, but my heart interrupts, tells me to let it all go. I breathe and take in the impermanent formlessness of life. The boat was smooth; now we're flopping; soon the ride will be soft again. I tap into a peaceful place within and grow curious to experience the reality of a stalled front. The boat tosses as I breathe and relax with a cup of tea, repeating to myself, "And this too. And this too. And this too."

25°04'38" N, 68°23'07" W // December 1, 2010
280 nautical miles northeast of Turks and Caicos

The big wind has stopped and left me crap to deal with—feeble coughs of air from random directions. Dave and I hoisted the mainsail before he went to bed and now it fills and sags indecisively. Gloom lingers. I've got *Wild Hair* pointed at a dime-sized patch of clear sky in the southeast, and she meanders aimlessly like a daydreaming child wandering home from school. Nothing is happening. My mind skims the indigo ocean.

"This here is the largest wilderness on the planet," I say, tour-guide style to my ship in an effort to stir things up. "And you might be interested in knowing lots of people have retreated to the wilderness when life got tough. Folks like the Buddha, Jesus, and more recently Henry David Thoreau. They wanted answers society didn't offer."

The sail flaps in the wind.

"All right, all right, all right. You're not impressed. I can tell you think wilderness is an iffy thing because humans have tinkered with every square inch of the planet or its atmosphere. Yet you can't deny that we're here right now in what poet Gary Snyder calls the 'primary temple of wilderness,' can you? Aren't you in your temple, *Wild Hair*?"

The hull sloshes indifferently. I must be losing my mind. My fingers unscrew the water bottle lying next to the seat and I take a swig. Actually, where exactly am I? What the hell is wilderness?

Some people think of wilderness as a godforsaken place of savagery and danger. Others see it as a garden that can flourish with human intervention. In a way, both of these ideas hold some truth. Indeed, entire professions exist—think engineering and medicine—that are aimed at eliminating or containing the wilderness and its organic processes (like the natural flow of water or the spread of disease). In general, more developed societies have been terribly prejudiced against indigenous peoples around the world who, for countless generations, have made their way in the wilderness in harmony with the land. Why would we shun people with this kind of knowledge? I don't get it. I take another drink and make a note of the time and *Wild Hair*'s position in the log book.

It seems to me that humankind's opinion of the wilderness fluctuates moment to moment, and is itself impermanent.

My mind lands on Frederick Jackson Turner and the *Frontier Thesis* he presented at the World Columbian Exposition in Chicago in the late 1800s. It was his idea that the wilderness was *the thing* that transmogrified aristocratic Europeans into rough-hewn Americans who valued personal strength and individuality. He said that basically a bunch of socialites from the other side of the pond got their butts kicked by forces that didn't give a fig about who their daddies were. The licking our forefathers took from the wilderness gave birth to a new spirit of democracy, the notion of small government, and America's unique identity. Yeah, I understand how time in the wilderness changes a person.

But then there were my professional predecessors and heroes (trained scientists who valued wild places as both the building blocks of a sustainable future and the antidote to civilization's ills) who watched as wild places started disappearing last century faster than gambling chips in Las Vegas. When I was five, the Wilderness Act was passed to set aside places "untrammeled by man." (A great word: untrammeled. The aspiration behind it is so pure. If only there were truly untrammeled places left.)

Sitting here on my ship in the middle of the wilderness, I don't believe wilderness is ultimately a garden to manage or a void to tame. Wilderness does its own thing, and it's up to me to jive with *its* natural law and order—not the other way around. The Great Atlantic Teacher is rearranging my sensibilities, my agendas, and there's not a damn thing I can do to force my will upon it. The reality that I'm not ultimately in charge makes me humble and cautious, attentive to what's happening, and grateful for every moment I am not in crisis. The wilderness doesn't particularly care who my society (or my technology) is as it transforms me into a new identity as a child of the Earth. I can feel it happening; I'm different from others in my tribe. I've become maladapted to civilized society, sailed closer to the workings of nature and farther from my culture.

Oh well. As Thay says, happiness doesn't come from adapting to an ill society. It's smart to distance myself from the flawed logic of dysfunction. I take a long pull of air into my lungs. It feels fresh and clean. For now, I'll keep searching out the divine wisdom contained in this ecosystem—the wilderness of water meeting air.

23°29′27″ N, 67°33′07″ W // December 2, 2010
235 nautical miles northeast of the closest land: Turks and Caicos

The world is still. *Wild Hair* floats like a dragonfly on a glass-smooth pond. Clouds reflect off the water.

"What direction is the wind coming from?" Dave asks. He is determined to suck forward progress from absolutely no wind. The indicator at the tip of the mast swings as the boat barely sways. This sends confusing signals to the electronics, making the gauge spin: swing—the wind is from port; swing—the wind is from starboard.

"There is no wind."

"No, look at the wind indicator. What does it say?"

"It says there is no wind." I have tried to tell him there's no point in fussing over the sails. But I can't argue with him when he gets like this. All I can do is wait until he's exhausted his options.

I for one absolutely love that we're becalmed. This is what happens to all the great sailors, historical and fictional; now we're among greats like Columbus and Ahab. For the hell of it, I contemplate mutiny for fear of falling off the edge of the world.

Wild Hair lacks the forward momentum necessary for steerage. Still, Dave spent the past two hours concocting strategies to lasso the sails in place. He's also locked the wheel hard over into a turn in effort to make the rudder hold a course. All this effort and still the sails hang limp and folded, flags on a windless day.

"I know what we can try." Dave steps toward the mast.

"I am so glad I wasn't in your operating room these past twenty years," I mutter.

"What's that?" He looks at me and registers for the first time that I think his efforts are pointless. "Do you want to stop?"

"Yes! You woke up rested and I'm still waiting to sleep and none of this is going to work. There's no wind." I go below to hear Chris Parker's broadcast before sleep and leave Dave to clean up lines.

The SSB signal hisses and pops; I figure the sun's magnetic field must really be in flux today.

"I'd like to start the broadcast by speaking with any boats sailing far offshore, over," I hear Chris Parker say.

"*Wild Hair, Wild Hair*," I say, microphone button pressed. Chris should answer—*Go ahead, Wild Hair.*

"Is there anyone offshore who wants to talk?" Chris asks the radio community a second time.

"*Wild Hair, Wild Hair.*" I say clear and loud.

"*Buffalo Gal, Buffalo Gal*," another boat chimes in. "Yeah, go ahead *Buffalo Gal.*"

"*Wild Hair* is trying to reach you, Chris. We can relay."

"OK great, *Buffalo Gal*. Please relay for *Wild Hair.*"

"*Wild Hair*, this is *Buffalo Gal*. Please go ahead and we'll relay for you."

"Roger that, *Buffalo Gal.*" I announce our position, the fact that we're becalmed, and ask if we should motor now or wait for the wind to arrive. *Buffalo Gal* repeats my message. Static and electronic tones interrupt the signal. Chris isn't catching *Buffalo Gal*'s entire message.

"*Buffalo Gal, Buffalo Gal*, this is *Whiskey Sour*. We can hear you and will relay your message from *Wild Hair* to Chris."

"That's great, *Whiskey Sour*," *Buffalo Gal* says. I hear *Buffalo Gal* repeat our message to *Whiskey Sour* and *Whiskey Sour* restate it to Chris. It's an impromptu pony express that makes me feel very far away. But the spirit of community, even all the way out here, is pretty neat. Again, Chris's answer is smudged by static. Other sailors jump into the relay, stepping on each other's voices. I ask *Buffalo Gal* to repeat, but they couldn't hear Chris or *Whiskey Sour.*

Unexpectedly, the radio goes silent; there's not a lick of chatter or static. In the void, Chris's voice says loud and clear, "Wait for the wind to reach you." I feel as if the heavens parted and God's voice landed upon me. Then a dozen sailors chime in telling me to wait.

"Roger, roger, I got it," I respond, and the radio bursts into a chorus of "Roger, roger—she got it Chris." The cacophony is at once funny and heartwarming.

"We have the day off," I say, swooping Dinghy into my arms.

"It's bacon day!" Dave says. "It's so calm, how about I cook up some pancakes?"

"I'll get out the coffeemaker and we can have brewed coffee, too."

After breakfast, Dave, Dinghy, and I collapse into a sleep that lasts all day. We wake for dinner at dusk and, because the entire universe is at peace, we sleep all night long, too.

In the wild blackness, *Wild Hair* drifts six miles south and west.

23°14′41″ N, 67°37′09″ W // December 3, 2010
229 nautical miles northeast of Turks and Caicos

"I'm not going to clip my tether to the jack line," Dave says the next morning.

"What? No way. You and I are going to follow our safety procedures at all times. I count on you to do that."

"Look at it out here. What's going to happen?" He has a point. This peaceful water feels like a northern Wisconsin pond; mounds gently lift and drop in lazy aquamarine swells. The world is visible for miles in all directions. "Clipping on and off is a pain. I'm sick of it."

"It doesn't matter. You know they say most accidents happen when things are calm." Dave pouts as he clips his carbineer to the line and walks forward. I clip on and walk aft to release the mainsail's preventer.

"Ugggha." I look toward the strangled sound of Dave falling on his back. The whisker pole slices inches above his flattened torso. The pole (a ten-foot aluminum stick used to suspend a corner of the forward sail in light wind) is like a tree harvested for timber dangling on a cable; *Wild Hair*'s mast—sixty feet in the air—is the crane supporting the cable. The pole soars thirty feet to port over the water, then cuts over Dave again as it swings thirty feet to starboard. I know what happened. Dave freed the pole from the sail and the rocking motion of the mast set the pole whipping. On the next pass, Dave lifts his hands and manages to grab the stick.

"Oh my god, honey. Did it hit you?" I move to him, clipping and unclipping.

"No. I saw it coming from the corner of my eye and dropped."

"Holy crap."

"Yeah, it could have knocked me unconscious and sent me overboard. Did you see that?"

"It was the scariest part of the trip so far—"

"And it happened in seconds in dead calm." We look at each other in stunned surprise.

At 07:20 in a light breeze *Wild Hair* is moving at five-and-a-half knots on a one-three-zero heading. I am no longer willing to declare this a beautiful day—I can't possibly know if that is true, given impermanence and all. It is a beautiful moment, however. In twenty minutes, or even a few seconds, disaster might strike. But this moment is bliss.

"I've got it," Dave shouts, lifting me from my thoughts. "I know our next business venture. What are those nautical-style clothing lines called?"

"You mean Nautica, Land's End, and . . . maybe Ralph Lauren?"

"Yeah, forget about them. Those people have obviously never even been sailing. We know what the real offshore sailor wears—underwear under the PFD."

"Oh my God," I laugh. "You're not ready for a catalogue."

"Think about it. It could be a new apparel line for the West Marine. We could call it 'Skipper's Secret.'"

"Yeah," I say. "For the nonsensual times in your life."

"I see vintage elastic and strategic stains of beef bouillon on pre-stretched cotton."

"It will signal to your partner 'don't bother coming hither.'"

"Right. And at the naughty bits, where others add feathers or beads, we will have scales."

22°11'00" N, 66°25'59" W // December 4, 2010
225 nautical miles north of Puerto Rico

Early this morning, I had a dream: I walked in tall, clean woods; the understory was soft and open. Branches sheltered me from the oppressiveness of an incessant sky. Everything was green and smelled of pine and earthworms. Birds sang. Chipmunks chittered in retort. I clung to the place as I woke, longed to stay in the cool shade to commune with trees. Here, the sun peeping at the horizon may paint clouds orange, red, and yellow, but green is nowhere to be seen. I crave green in this blue-gray world like a teenage girl craves her first kiss. But I woke and remembered my separation from the green earth is temporary. Soon I'll be with forests, and then, in the inevitable churning of reality, this restless spirit will probably crave the ocean blue.

During my next nap, I dreamt I was eating an almond scone at Starbucks. It was just how I like it: moist and not too sweet. I woke up and ate a stale biscotti.

Chris Parker said today is the day to point east. It's our last opportunity to travel the remaining eighty-five nautical miles to I-65. On my watch, I'm pushing *Wild Hair* hard and fast into rough waves. The sky and water are moody and confused. For miles in every direction, white-caps collide. A seventeen-knot blow moans across the canvas and waves growl down the hull. Clouds—like glued clumps of torn fibrous paper in various colors of charcoal—hang; their full bellies are low, inches above the mast. The turquoise sky between the clouds is so clear it's sharp. Negative and positive shapes of sky and cloud lunge back and forth like the figure ground reversal of an Escher painting. My skin registers—on an animal level—the evaporation giving rise to these giants. Somewhere, great quantities of airborne water will drop, reviving thirsty fields and bathing homes. Where exactly is the ocean going, I wonder.

Wild Hair humps east through a raucous, liquid mess.

We couldn't know what the journey would be like before we did it. Books, bar conversations, and lectures at sailboat conventions provided clues, but didn't come close to revealing what it would be like. We had to travel into the wilderness and find out for ourselves.

Still, I can hardly comprehend this world gone wide, the complexity of the elements at play. Does my being here matter to this place as much as it does to me? Probably not. I am like a Shakespearian tertiary charac-ter who passes comically through a scene without affecting the plot. But jesters matter: Shakespeare valued the distraction, and perhaps Earth does as well.

East. *Wild Hair* pulls east.

Dave and I grew into technically proficient sailors on this journey. Our novice state of only twelve days ago was impermanent. A skilled response to changes in wind direction is deeply ingrained now, second nature. I don't shrink with intimidation when nasty conditions arise; I act to get ahead of trouble. I discovered stability in the midst of instabil-ity, and found refuge in the strength of our boat, her crew, and our skills.

Slam—goes *Wild Hair's* bow into the next wave. Rivulets of water squirt skyward, like a squeezed orange. *Slam . . . Slam . . .* East . . . East.

I am asleep when Dave crosses 65 degrees west longitude at 23:00 and makes the final turn south toward our destination. When I wake, the boat is stretching long through the ocean and feeling as silent and settled as a yoga master. Warm and steady trade winds pack our sails.

"Were paparazzi snapping pictures at 65 degrees west to capture the moment?"

"No, there were no photographers. It didn't even look like an intersection. It was pretty anti-climactic."

"Did you snap a selfie?" He shakes his head no.

I look east over the port rail. "Well hello, tradewinds. My, you are beautiful today."

20°58'10" N, 64°56'36" W // December 5, 2010
160 nautical miles north of St Thomas, USVI

"We may arrive in St. Thomas tomorrow morning," Dave says as I wake. "It's time to burn the Skipper's Secret and break out the bikinis."

"Before now, I was afraid to hope," I confess.

"Tell me something," Dave says, as I get ready for my watch. "Have you . . . have you heard voices?"

"Heck yes! I can't believe all the people that are here. The boat's crowded!"

"Oh good. I thought I was going a bit goofy."

"I don't think so. Who did you hear?"

"Everyone," he shrugs. "And they aren't just," he searches for words, "faint whispers. Right?"

"No." I look straight at him. "People and voices are all here. They're not inside my head; they are in the room."

"OK. That's good. If we're crazy, at least we're crazy together. They can put us in the same padded cell."

"I'd like that. Just think what we could do in a padded cell!"

Land. Tomorrow morning. It seems impossible. I imagine a long, hot shower, salad with crisp lettuce and tender tomatoes, and walking Dinghy on her leash for more than four steps. I will sleep in a quiet bed instead of a vaulting, three-sided berth. I will go places without holding on or chafing in a life jacket. I'll window shop and people watch to absorb culture and the energy of others. The kindness of a stranger will feel tender as we help each other through the day. I will wear clean and pretty things and eat food prepared for me by another person without worrying about the contents of my plate flying through the air. How much pineapple can my umbrella drink (with ice!) hold? I'll slide a chair out from under a table before sitting down and relish movable furniture that doesn't have to be screwed to floors and walls. Maybe I'll even swim in this beautiful, warm water without fear of drifting away?

I spot my first Frigatebird, a harbinger that we're close to land. I feel comforted.

These weeks with Dave were precious. His energy was unlimited. He was calm and kind under stress, able to voice our common goal, keep us focused as a team. We took care of each other by making certain the other stayed hydrated and we nursed each other's wounds. We took care of the boat as it flip-flopped between states of repair and disrepair. We split responsibilities and played to each other's strengths. I managed communications and Dave oversaw fuel and power usage. I made meals and Dave cleaned up. I let him decide the amount of sail we'd fly and he listened to me on how the sails should be trimmed.

Trade winds stop blowing at 14:00—just as Chris predicted. We are becalmed 152 nautical miles north of St. Thomas. But there is enough diesel in reserve to drive to our destination. I start the engine for the final leg, we drop the sails, and *Wild Hair* becomes a motorboat. Sailing is not sailing; therefore, it is sailing.

REFLECTIONS ON IMPERMANENCE

18°35'47" N, 65°03'56" W // December 6, 2010
24 nautical miles north of St. Thomas, US Virgin Islands

It's dawn and I am alone at the helm. The lights of St. Thomas sparkle gold and red in a flat, wide cluster. I am uneasy because I feel I don't fit in with ordinary society anymore. The ocean made me different. It beat and bullied me, teased and seduced me, shouted and whispered and altogether exhausted me until I cracked open, let go, and began appreciating not patterns in the wilderness but the ultimate formlessness, unknowability, and impermanence of the sea (and everything else in the universe for that matter). I'm completely and utterly ruined for good company, maladapted to be with anyone who actually believes what they see, feel, and think and who depends on things to continue as they are.

People who expect the sweet spot of their world to last forever make themselves miserable. Rather than surfing waves of becoming and ceasing, they cling to their preferences and experience epic loss when conditions inevitably change. They trust their perceptions absolutely and find fault when life shifts. They spin stories, assign blame, and develop prejudices. They expend vast resources resisting change and even, collectively, build large militaries to make war. Before the Great Atlantic Teacher hammered the lesson of impermanence deep into me, I was someone who clung to my perceptions and preferences like battle flags. I can't be like that anymore.

In the past few weeks I've seen weather forecasts turn on a dime; the wind direction endlessly clock north-south-north; our bodies injured and well, filthy and clean, sleepy and awake; and a joyous holiday turn to heartbreak. Storms blew over. *Wild Hair* zig-zagged between soundness and breakdowns, and sails trimmed perfectly became dangerous. Dave and I have been both alone and together with family living and dead. And all the while, waves of emotion—from happiness to mutinous anger to transcendent endurance—rebounded between my heart and mind. I felt as if I were back in Chicago growing up, when cab drivers liked to say, "Hey, if you don't like the weather, wait a few minutes." There's profound wisdom in these words.

Impermanence is a precious insight: like my boat, it helps me surf the waves of change and avoid getting swamped when they arrive. There's value in remembering the unique preciousness of this moment without clinging to it. Contemplating the impermanence of a thing or moment isn't morbid—growth and the potential for transformation are rooted in the reality of change. If the world stayed just as it is, then there'd be no such thing as possibilities, learning, or justice. I want to keep the experience of impermanence alive and live fully from this place of insight, even when I'm on land and my days feel ordinary and routine. Perhaps when I sit in meditation I can reflect upon how things around me change: thoughts, trends, relationships, and this aging body. Maybe I can touch impermanence when I read reports about shifts in Earth's climate. I can also spend time outdoors and intentionally find evidence of seasons in transition in the migration of birds.

All at once, the bright moon escapes from behind a cloud, and *Wild Hair* and I glow in blue light.

"Who am I kidding? Moon, *you* are my friend. Knowing you has severed me from my species and I have no interest in going back. I want to be with you." The moon doesn't flinch but listens deeply to my words.

I am maladapted to humankind. If I reimmerse myself in a civilization convinced it will go on indefinitely just as it is, if I mingle with people who cling to fixed views of right and wrong, us and them, have and have not, wealth and poverty, safe and unsafe, I'll forget the lesson of the Great Atlantic Teacher: that absolutely *everything* is in flux and nothing is solid. Distracted by society's trappings, I'll get complacent; my thoughts will grow small and selfish and my body comfortable and flabby. I'll slip into my old ways of being and once again become a driven, frustrated, exhausted woman determined to *change* the world without really *knowing* the world.

This thought makes my hide itch.

Screw it. I can't go back. I'm going to continue sailing, living in the present moment, experiencing the Atlantic, riding endless waves of impermanent cloud patterns, sea states, sounds, thoughts, feelings, and ideas. Life is clear out here. What's broken, Dave and I fix. What's pleasant, we relish. What's unpleasant, we wait out. What's dangerous, we amend. I'm going to keep letting go of ideas around well-being and derail my stubborn determination for things to be just so. I'll stay free of the limitations of form and perceptions and live from a place of transcendent endurance.

"Atlantic, beloved ocean, *you* are my true home. I will veer off and keep sailing. Won't Dave be surprised?" A passing breeze wraps my shoulders hug-like as it moves along.

Suddenly I catch myself and take a breath. I realize I'm not welcoming the impermanence of the voyage but clinging (once again) to my preferences. I *do* in this moment like life here better than land. But I ought to be looking past my fleeting conceptualizations about both the sea and society. After all, my natural habitat is land; my community is people.

I breathe again and try to let my emotions and judgments about returning rise, exist, and disappear.

Closer now to St. Thomas, I can see the lights on shore illuminate roads and buildings. An insight begins to percolate: the lights themselves are evidence of love—people laboring to offer kindness to one another. Society is imperfect but beautiful. We teach our children, care for our sick and elderly, protect the environment, feed those who are hungry, create peace in the hearts of those who are troubled, and work for justice. There is formlessness in society. Yes, many people cling to their misperceptions and behave badly, but these actions are impermanent and there's no need for them to ultimately define our species. Humankind is extraordinary and has given Earth the gifts of literature, medicine, dance, science, and education. We make each other laugh for the joy of laughing. We create beautiful architecture and plant trees for future generations. As a species, we are both wearisome and inspiring, and the complexity makes the essence of our kind ultimately fluid, unknowable, and beyond description—like Earth herself. Society is not society, therefore it is society.

Feeling this truth echo from my baseball cap to my stocking feet it dawns on me that society can turn on a dime. We are capable of greatness, and anything is possible because of the nature of impermanence. I breathe for several minutes. *Wild Hair* hums evenly through gentle waves.

There's no difference between people and Nature. We are born of Earth; we are the stuff of Earth; and we will return to Earth. Humankind is impermanent. But somehow this realization isn't sad or morbid. There is joy in the knowing. We're precious and fleeting—like the sweet smell of a baby's breath, like fresh cut flowers on a wedding day. And when we're gone, something else miraculous (and equally fleeting) will take our place in a great continuum of arising and ceasing.

"Not so fast, Heather," I mumble, smiling. "Let's not jump ahead to our demise. Stick with the fact that humanity is capable of greatness." The thought turns in my imagination. I decide to nourish myself in the transition back to a life on land, by paying attention to the evidence of humanity's love, finding it everywhere.

"This is a good way to be in a suffering world, dear Atlantic Teacher," I say with energy into the night. My shoulders lighten. A smile grows on my lips. Now I know where to look to counterbalance the chaos of petty ways. In the turmoil of impermanence, in the confusing array of mistakes and possibilities, humanity's love can be the salve for my pain. I stand and test out what feels like new legs wobbling on the threshold between an ungraspable ocean and an inscrutable civilization. As I stretch, I smile to the moon. "You'll be with me on land too, won't you?" A wave splashes through the open cockpit panel and salty drops land on my cheek.

It is said there are two kinds of sailors: those who sailed offshore and those who want to. Today, Dave and I graduated into the first category. We completed the rite of passage from the transmission of learning to proficiency, from unknowing to knowing, from theory to experience. I realize now I sailed not away from society but deeply *into* it. All that arises ceases to be (this is the reality of impermanence). The land is the ocean and the ocean is the land (this is the reality of formlessness). Taken together, we cannot ultimately know what's going on or what lies in store. All we can do is surf the waves.

At the helm of my ship, blanketed by the moon, I sit and breathe in peace.

~~~

Some months later, to celebrate the insight of impermanence, I wrote this poem. I recite these words whenever I need to remember how to surf reality and touch the universe of possibilities.

The whole
formless and in flux
becoming and ceasing
precious and fleeting

perceptions you are flawed
body you will not continue
civilization you will end
beloved Earth, we will leave you.

My darling let us dare
to be great
surf and transcend for
love prevails

9 MINDFULNESS

It's 02:00. The night is magic-marker black. The stars and moon have abandoned *Wild Hair* and me and left us to grope in the dark. Bug-eyed, I scour the surroundings 360 degrees, but there's only one light bobbing a half-mile ahead. It's the stern bulb on a home-built tri-hull named *Extreme*. Our new friends Bob and Jillian are the crew. Together, we're traveling across the deepest channel in the eastern Caribbean: the Anegada Passage.

Dave, Dinghy and I hit the cruiser's jackpot these past few months. After flying to Phoenix to celebrate Christmas at my parents' home and hopping to LA to visit with Eland and his new girlfriend, we returned to gunkhole aimlessly through the Virgin Islands (US, Spanish, and British). *Wild Hair* took off whichever way our whims dictated. My craving for green was satiated hiking forested trails that ended at water-falls whose rocks were etched with ancient petroglyphs. Dave and I meandered hand-in-hand through historic towns where just below the surface of glitzy hotels and beach bars reverberated remnants of the brutal energy of pirates and slave masters. The two of us ogled over posh houses with delicious looking swimming pools dangling above drop-dead ocean views. We shopped at markets for avocados the size of my head and fruits and vegetables the likes of which I'd never seen. We toured plantations, visited an artist colony and the workshop of a mahogany craftsman. At dusk, we watched classic movies on the beach projected onto a canvas stretched between palm trees. We were flabbergasted by the energy of the equatorial plant kingdom in a botanical garden on Saint Croix—a climax of evolution I could not have fathomed while in the middle of the ocean.

And oh, the Caribbean water! Dave and I snorkeled one coral reef after another. We dropped anchor and swam in the company of endangered hawksbill turtles, spotted eagle rays, a battalion of squid, and even an octopus (who did her best to stay hidden from me by meticulously color-coordinating her outfit to the surroundings). We body surfed in breaking waves on a beach legendary for its natural beauty. On another day, the two of us swam in and out of caves with scores of Blue Tang Fish. When Maggie came to the Caribbean for her spring break vacation, she didn't care too much about the details of sailing, but she was all about anything that had to do with the Caribbean and her creatures.

Dinghy loves this life. At least once a day she got shore leave to explore the latest island from her vantage at the end of a leash. Her chest swelled and she stepped into confrontations with small dogs and wild iguanas. (Who knows what she would have done if we didn't keep her on the line.) Wherever we went, people snapped photos of a cat on a leash to send to their daughters or post on their Facebook pages. I believe images of Dinghy are wallpapering people's cellphones around the world. Once, arriving to a new marina on the opposite side of the island we had been visiting, the fellow catching our lines said, "You're the people with the cat who walks on a leash. I heard about you!"

Of course, our adventuring was regularly interrupted by the necessary repairs of the latest thing that broke on our dear ship.

Dave and I met Bob and Jillian aboard *Extreme* in the cozy harbor of Isla de Culebra in the Spanish Virgins months ago. Our paths crossed many times since. Actually, we've come to know them and the gaggle of boats and boaters they caravanned with down the island chain from the United States to here. (Rather than going offshore, the bunch took the slow route island hopping south and seemed to have a jolly good time doing it.) But tonight, our two boats are leaving the Virgin Islands and starting the trek south to arrive in Grenada (below the line of the hurricane zone) by May 30th—the end of the sailing season. *Wild Hair's* mainsail is down and the motor hums. Our anchor let go of British Virgin Island sand at sunset and will dig into French earth later this afternoon in St. Martin. It's an 18-hour, 75-mile journey to the French West Indies.

Mariners nicknamed this corner of the ocean the Cape Horn of the Caribbean. It's rough going here, where the Atlantic slams into the Caribbean. Great ocean swells surge from the north, trade winds blow from the east, and currents push from the west. In the midst of these opposing forces, *Wild Hair* tips up and down and whirls right and left in erratic combinations. All this whiplashing and corkscrewing makes me queasy. I swallow hard to keep food in my stomach.

"Hey you out there. Ocean, my friend, please give me a break. Can you at least give me a view of the horizon?" It's no use. My gut convulses and my dinner launches into a two-and-a-half-quart saucepan readied at my side. I wipe my mouth on the back of my hand, take a breath, and continue steering.

I've got on my sweatpants, t-shirt, and rain jacket to buffer the ocean breeze. A minor burst of fresh tropical air revives me a bit. My innards calm. It really is a lovely night; the waves are just so odd. I press on and let Dave and Dinghy sleep.

Chris Parker said tonight is the only decent night in the next several weeks to cross the Anegada. Air currents are topping off at fourteen knots; waves are piling to five feet. These conditions are not the six-knot breeze and one-foot seas I prefer to point straight into, but this is the Anegada. I'm willing to take what I can get, because ideal conditions just don't exist this time of year.

Night blind, I cruise along feeling the soundness of *Wild Hair* as she moves through disorganized sea. A shooting star zips at the edge of my sight. I turn, but it's gone. The glimpse strikes me as somehow a gift and a missed opportunity. I shake off my sentimentality before I turn melancholy.

Wild Hair's VHF radio is crackling with conversation. From what I understand, a freighter carrying fragile produce lost power and the captain is desperate for refrigeration and a tow. For some reason the US Coast Guard in Puerto Rico can't hear the distress call, so Bob and Jillian are taking turns relaying messages like old-time telephone operators. During a break, another voice—one thick with an accent—jumps in.

"All vessels, all vessels, all vessels, this is *Breathless, Breathless, Breathless*. We are looking to speak with any boats crossing the Anegada tonight. Come back." I jump in.

"*Breathless, Breathless, Breathless*, this is *Wild Hair, Wild Hair*. Please answer and switch to channel 68."

"Roger, 68," the woman says. I change frequencies to a conversation channel. "*Wild Hair, Wild Hair* this is *Breathless*. I just wanted to say 'hello' and make you aware of our presence." As the conversation unfolds, I learn *Breathless* is twelve miles behind *Wild Hair*—about two hours away, should there be trouble. The woman and her mate are from the north of England. But they bought their boat in the United States and sailed a route similar to our own. It's funny that we've not crossed paths before.

"It's dark out here, no?" she asks.

"Yes. It feels like sailing in a cave."

"I can't remember a night so black."

"No."

"Well, I just wanted to connect. Let's carry on."

"Keep in touch if anything comes up. Have a good sail. *Wild Hair* out."

The autopilot is doing a good job, so I am an underemployed supervisor. I entertain myself by practicing mindfulness. Breathing in, I know I am breathing in. Breathing out, I know I am breathing out. In . . . Out.

I take many mindful breaths, paying attention to the sensation of cool air passing through my nostrils. The bow vaults as *Wild Hair*'s stern shoves right. My mindfulness shifts to the sensations of my body—the way my spine absorbs the motion. Breathing in, I notice the sensations of my body. Breathing out, I notice my body. The boat whirls and my stomach lurches. Breathing in, I feel my body respond to the mixing sea. Breathing out, I smile to the ocean wilderness. In, mixing . . . Out, smiling . . . .

All at once, the saltwater curling the port and starboard hull ignites in a Fourth of July sparkler display. The spectacle shocks my eyes—the glitter is like countless flashbulbs popping. We must have sailed into a red tide—a dense bloom of dinoflagellates or single-celled, plant-like creatures blinking with bioluminescence. As *Wild Hair* pushes and deforms their orange bodies, they complain by triggering a blue-white light that lasts a fraction of a second. Marine biologists suspect this reaction is a defense mechanism: flashing, they say, either startles predators away or attracts more to the area to threaten the first.

But it's possible these aren't dinoflagellates at all. 90 percent of marine life has the know-how and technical where-with-all to make blue-white light, so this spectacle could be bacteria, algae, worms, jellyfish, krill, shrimp, or any of a number of fish. The sparks continue. I sit in awe. I wonder if others sailed into the bloom.

"*Extreme*, *Extreme*, and *Breathless*, *Breathless*, this is *Wild Hair*, Check out your hull. Do you see bioluminescence? Over."

"*Wild Hair*, this is *Extreme*. No—we're dark over here."

"*Wild Hair*, *Breathless* here. We're not seeing anything either. But we'll keep an eye out. Thanks for the head's up. We're standing by on one-six."

I must be in the midst of millions of dinoflagellates, because they are the most common flashers on the ocean's surface. The show blazes all around the boat. If I stood at a distance and looked down, I'll bet *Wild Hair* would appear to ride on a stream of magic dust through the void of space. And who's to say what's really going on?

I think it is the mystery and wildness of the present moment that the Buddha wanted us to wholeheartedly embrace through mindful awareness. The Great Atlantic Teacher is driving home the Buddha's point by sensitizing me, on and off the boat, to a limitless and divine wisdom permeating the world here and now. I practice being curious about the little intimacies of my own experience with organic processes. I look for lessons, moment-to-moment, on the natural order of things. I discover the Earth is teaching all the time in clear and precise ways, and she is revealing important truths I cannot reason my way toward.

When slivers of light launch from the sea and skip the tips of the waves for impossible distances, flying fish show me the how and why of liberating myself from everyday boundaries of misperception.

I notice compassion gripping my chest when, during the washing of the ship's deck, a bulky manatee rolls in the fresh water pouring from my garden hose. I see scars striping her body, poorly healed cuts made by motorboat propellers. Open and unafraid of humans, the manatee teaches me a deep lesson on fearlessness and tolerance.

When seagulls circle in currents of wind, and splash headlong into the boat's wake to feed on creatures stirred to the surface by our propeller, I am swept into the aerial acrobatics. Death is found in every life and life is found in every death. Life and death become one.

My silent inter-species conversations with dolphins when they come alongside the cockpit to look me steadily in the eye and say hello makes me understand that there is nothing unique or special about my human knowledge.

When sharks circle in the water in which Dave and I swim, I instantly see the most fundamental workings of ecological systems, experience what it is like to be demoted a notch on the food chain, and wake up to my own impermanent nature.

I am cultivating ecological mindfulness as I go. Everywhere I look, things animate and inanimate have something to teach. The Great Atlantic Teacher is showing me divine wisdom permeates the whole. This discovery is revolutionary to me.

"Dinoflagellates, I see you, and I'm sorry to be messing with you," I say. I imagine a single dinoflagellate floating peacefully with its kin, when suddenly *Wild Hair*'s hull sends it spinning. Frightened, it jolts, spending precious energy in self-defense. I feel a little guilty disturbing the dinoflagellate society, but—honestly—I'm mostly happy to have discovered their talents in the night. I'm glad they've made themselves known.

"Hey, you are not alone, Bozo," the creatures seem to say. "Be careful!" Now there's a teaching for modern times.

After many minutes, the flashing ebbs and then stops. There's nothing left to do but return my thoughts to meditation. I feel the sensation of my lungs breathing, my body sitting on the firm cushion, my hips, waist, back, and neck counter-balancing the motion of the boat's rolling. My gut stays calm. My mind is at peace.

A blip appears on the radar. With every rotation of the arm, a yellow speck inches a little closer to the center of the black monitor. Something is drawing near. The Anegada Passage is clear of buoys and islands, and recreational boats travel in an east-west direction, so I suspect this south-to-north moving object is a container ship. The radar tells me it's about nine nautical miles off. Is this the produce vessel I've been hearing so much about? Is it on the move again? I look in the general direction of the blip and, indeed, a cluster of red and white lights hover far away.

My instinct says I should let up on the throttle and slow *Wild Hair* down. I'm in no hurry to get to St. Martin. It's no big deal to alter my speed and let whatever this is cross ahead of my path. And yet our buddy-boat *Extreme* has a sophisticated technology I'd like to check out: an Automatic Identification System, or AIS. It would be great to see the technology in action. Besides, I trust Bob and Jillian—they're licensed captains who ran a charter business for years. Under the circumstances, I think I'll follow their lead.

Traveling with AIS is a sweet deal. It's like having a telepathic mathematician as a copilot. The instrument sends and receives the names, positions, origins, destinations, speeds, and bearings of neighboring ships. It anticipates collisions, sets off cautionary alarms, and calculates time and distance until impact. The big boys—vessels traveling internationally and weighing three hundred tons or more—are required to have AIS technology, but very few recreational boaters install it because of the price tag. The problem is, commercial vessels sometimes ignore recreational boats like hogs ignore fleas, even when we radio them with something important to talk about (like a plan to avoid collision). With AIS, a small boat can hail a commercial carrier *by name*. The larger ship is legally bound to respond when someone—anyone—hails them by name. This is how AIS technology can increase safety at sea.

"*Generate, Generate, Generate*, this is *Extreme, Extreme, Extreme*," Bob calls the ship's radio operator. "We are a small sailboat traveling eastbound across your path in tandem with a second sailing vessel one mile to our south and west. Do you pick up two vessels on your radar? Over."

"Roger that *Extreme*, this is *Generate*." The man's voice grates with tobacco undertones. "I see two boats traveling zero-nine-five degrees at approximately five knots. We will be monitoring your course and speed. Over."

"*Wild Hair, Wild Hair, Wild Hair*, this is *Extreme, Extreme*. We picked up a commercial freighter on our AIS traveling from Antigua to San Juan, Puerto Rico with a bearing of three-three-zero degrees and a speed of eighteen knots. They will cross our path forward of our bows."

"Roger, *Extreme. Wild Hair* is standing by on one-six." This is so cool. No longer are we anonymous strangers, the proverbial ships passing in the night. I wonder, should I establish contact with *Generate* myself? No. They've got us in their sights and that's enough.

The radar blip closes in and my fingers twitch to let up on the throttle.

"*Extreme, Extreme, Extreme*, this is *Wild Hair*. What is our closest point of approach to *Generate*?"

"Uh, hold on," Bob mutters. Instantly, an alarm onboard *Extreme* starts buzzing caution, and the sound reaches me across radio waves. "That's a very good question, *Wild Hair*," Bob laughs. "Let's reduce our speed to three knots and give *Generate* some room to pass."

"Roger that, *Extreme*, *Wild Hair* is slowing to three knots, over." I reduce power and look to the right, but I can't even see *Generate's* lights any more. Where did they go?

"*Extreme, Extreme, Extreme*, this is *Generate*. We are maintaining our course and speed but are concerned with your movements. What are your intentions? *Generate*, over." The yellow blip continues to near the center screen of the radar.

"*Generate, Generate*, this is *Extreme*. The two boats you are tracking slowed to three knots to accommodate your course. Over."

"Uh, yeah. *Extreme*. This is *Generate*. Would you like us to take evasive action and change course to avoid collision?"

At once, daylight explodes from above. A spotlight mounted ten stories overhead illuminates not only *Wild Hair* but also an impenetrable curtain of red steel fifty yards ahead. Sailing vessel *Extreme* splits the difference between *Wild Hair* and the wall.

"Shit!" I'm on my feet, spinning the wheel hard to port; *Wild Hair's* bow swings. I yank the throttle into neutral and check on *Extreme*. It too is coming to a standstill and avoiding collision. My heartbeat thuds in my ears as I watch a wall of rivets, rust, and peeling paint glide past my view. The mass has no beginning, top, or end—there is only an endless, red middle. *Generate* is solid, escalator-smooth. She is a dry bulk cargo ship as long as a city block and weighing nearly two-hundred-thousand metric tons. Her spotlight remains concentrated on our smaller boats, an effective command to stay put.

Dave climbs halfway up the companionway ladder and looks forward. No doubt the change in engine pitch and sudden daylight woke him.

"Huh," he says.

"Yup," I say. Without another word, he goes downstairs and back to bed.

After many minutes, *Generate* clears our path and extinguishes daylight. The yellow blip on my radar is left of center now. My hands shake as I go about putting *Wild Hair* back on course.

"Rule one," I remember my dad saying, teaching me how to drive when I was a teen. "Never get in anybody's way. No one should ever have to step on their brake or speed up because of you. Drive like that and you won't crash."

I nearly violated the first rule of driving tonight.

After several moments, Bob calls: "*Generate, Generate, Generate,* this is *Extreme, Extreme, Extreme.* Thank you for your assistance. It has been a pleasure speaking with you tonight. Have a safe passage. Over."

"*Extreme,* this is *Generate.* Roger that. You have good night and a safe passage as well. Over."

I feel heat flushing my face. Pitting my twelve-ton fiberglass sloop against a steel ship fourteen thousand times larger was a damn fool thing to do. I want to make light of our brush with disaster. Maybe I should radio, "*Generate,* this is *Wild Hair* traveling in tandem with *Extreme.* Thank you for shining a light on the situation. Our close inspection of your portside hull is now complete. You are clear to go. Good night."

Ugh—that may be cute, but it's not funny. Under the circumstances, I think I'll just keep quiet.

How could a near-collision happen under my mindful watch? I wasn't zoned out or not paying attention. I knew full well that the ship was there. Why didn't I follow my instincts and slow down *Wild Hair* when I knew it was the right thing to do? Did I put too much faith in Bob? Did I fail to act because he was unconcerned? Maybe Bob and I suffered from groupthink—did each of us assume the other was responsible for a judgment call when in fact no one was leading? Or maybe I put too much stock in advertisers' promises that AIS would keep me safe. I knew better than to let this happen. Did I disregard my own knowing—abdicate my personal responsibility—because my faith in technology was stronger than my awareness of reality?

Frankly. I don't like my answers to any of these questions.

"Heather," I mutter aloud, "have you learned nothing from all these years of mindfulness practice?" My brow is knit, my mind bewildered.

REFLECTIONS ON MINDFULNESS

Thich Nhat Hanh declares mindfulness not a religion, but a miracle. Practicing mindfulness—he says—is the most important thing we can do. We do it "to know what's going on, not only here, but there."[30]

Mindfulness is a tool to connect our mind with our body. This in turn builds powers of concentration and insight. When we practice mindfulness, we experience the world more deeply. Time slows. Life grows richer. Sometimes the experience is pleasant. Sometimes, when we are troubled or hurting, it takes courage to mindfully look into the roots of our pain. And, now and then, the practice of mindfulness is neutral and we feel neither pleasant nor unpleasant feelings. Mindfulness is being aware and accepting of what is.

Mindfulness is a way of life. We are invited to practice it not only in peaceful solitude, but in every moment of daily living. We are aware of our sitting, standing, walking, talking, and eating. A red traffic light, ringing telephone, crying baby, or bird in flight are "bells" of mindfulness inviting us to come back to our awareness and reconnect our mind and body in order to pay attention to what is happening in the reality of the present moment.

When I practice mindfulness, my body and thoughts come alive in my consciousness and my understanding about what's actually going on deepens. When I shine the light of awareness on my environment, I expand my field of consciousness to take in not only what's going on in here but also what's happening out there. In doing so, I witness Earth as a dynamic, living organism. The Great Atlantic Teacher becomes my blood and tears. Earth's atmosphere fills my lungs and blood vessels. The planet's radiant heat becomes my warmth and energy. Earth's harvest stuffs my belly and turns into my muscles.

There's something more to the art of being mindful about the environment that the Great Atlantic Teacher made clear the night I crossed the Anegada Passage. It's not enough for me to just sit in a bubble and be deeply mindful of my interconnection with Earth when I know that danger is on the horizon, barreling down, about to strike. There's a name for the behavior of staying peaceful when urgent and decisive action is what's called for: it's called taking a "spiritual bypass" (taking the road that skirts the heart of the matter). That kind of behavior is dangerous. The lesson I heard loud and clear from the Great Atlantic Teacher was that I need to keep track of what's happening *in here* and *out there* and *apply* what I know in order to stay safe.

It's no wonder Thay has said time and again that mindfulness is a constant practice and not a permanent state of being, something I am to do all through my waking hours. The safety and quality of my life and the lives of those around me depends upon my broad and penetrating awareness—at sea and on land. It also depends upon my being aware of the consequences of my own inaction. No longer can I take my lead from those who appear unconcerned or indiscriminately *assume* people with expertise will keep me safe. I cannot abdicate my personal responsibility to others, especially when I know there are simple things I can do to keep myself and others safe. Becoming mindful of my environment means I can no longer put absolute faith in anyone (scientists, politicians, commercial interests, or technology—*especially* technology). I have to be mindful of my personal responsibility—both to myself and to my environment.

In the Anegada Passage, the Great Atlantic Teacher gave me a lesson that will make me a better skipper, but the teaching also resonates with the central question I carry with me as I sail: how am I to live in a suffering world? The answer arising from this lesson: pay diligent attention and *apply* what I know as I sail and as I engage with a world in trouble.

I think there's such a thing too as Ecological Mindfulness—an intentional consciousness raising about the dangers and suffering of Earth's systems and life forms. To my mind, Ecological Mindfulness *is* climate action because it seamlessly connects my human mind, heart, and environment, so that I might have the insight to know how to be and what to do to protect the planet.

One spiritual leader—the Venerable Bhikkhu Bodhi—practiced Ecological Mindfulness when he peeled away layers of reality to see the underlying causes and conditions of suffering that once were hidden. He identified Earth's most pressing global threats as: overpopulation, poverty, climate change, and biodiversity loss.[31] Then, looking beneath these dangers, he named societal actions perpetuating the global threats— actions like our consumption of fossil fuels, the domination of corporations and their demand for short-term profits, eating meat, and society's overconsumption of material goods. Underlying these harmful actions, Bhikkhu Bodhi shined a light on dangerous feelings and ideas: greed, fear, arrogance, ignorance, belief in discrete self-identities, misguided ethics, and the pursuit of infinite growth above other values.

Practicing Ecological Mindfulness, I consciously carry the planet with me as I sit, stand, and lie down. During mindful walking, I know I am Earth walking. Broadening my focus and putting my awareness on the difficulties of Earth and the ways I respond to Earth's challenges (or not), I can wake myself up to the moments when I hand off my autonomy or convince myself that others are either unconcerned or more competent in dealing with the problem. I can begin to understand the mechanisms at work that tip me into the semi-conscious groupthink.

None of us will practice mindfulness perfectly, but it is becoming more and more accessible, and there are so many opportunities to learn and practice. Already mindfulness is finding its way into schools, the halls of medicine, prisons, and businesses. It's time to apply the time-tested tools of mindfulness practice to consciousness raising about the difficulties of Earth. If each of us could cultivate Ecological Mindfulness and apply inner knowing and divine wisdom to the simple, direct, and compassionate actions within our reach, we would likely transform the world. Earth could heal and flourish.

It is in this spirit that I offer the following meditation as a training in Ecological Mindfulness.

Breathing in, I see the reality of ecological suffering on Earth.
Breathing out, I extend compassion to ease suffering.
In, ecological reality.
Out, compassion.
(Ten breaths)

Breathing in, I look deeply into the roots of ecological
    suffering (greed, fear, ignorance).
Breathing out, I smile to the roots of ecological suffering.
In, roots of ecological suffering.
Out, smiling.
(Ten breaths)

Breathing in, I see a path leading away from suffering
    (a path of insight, love, compassion, ethics,
    justice, moderation, and simplicity).
Breathing out, I see my actions contributing to the path of healing.
In, a way out of suffering.
Out, I contribute to healing.
(Ten breaths)

Breathing in, I cultivate reverence for both Earth
    and earthlings—without discrimination.
Breathing out, I free myself from fear.
In, reverence for Earth and earthlings.
Out, fearlessness.
(Ten breaths)

ST. LUCIA

VIEUX FORT

ST. LUCIA TO BEQUIA

ST. VINCENT

PORT
ELIZABETH BEQUIA

10 FIERCE COMPASSION

It amazes me we don't need a license to have this much fun. These past three weeks, Dave, Dinghy, and I hopped south down the Caribbean Island chain, relishing the legendary trade winds that blew steady as a warming convection oven. We dropped anchor in a new harbor every night: under a small peak on St. Martin nicknamed Witches Tit (whose likeness to the real deal was uncanny—or so I imagine); in the shark-infested waters of St. Barts; next to a field of goats on St. Kitts; past the sulfur-stink of Montserrat's belching volcano to Deshaies, Guadeloupe, where the local baker rowed out to greet us when we arrived to take our order for fresh baked croissants and a loaf of bread he promised to deliver in the morning. Did I want plain or chocolate croissants? Let me see.

Dave, Dinghy, and I wandered the streets of Terre-de-Haut on Les Saints and got soaked like mops in a bucket before treating ourselves to a multi-course lunch of French cuisine (while Dinghy waited patiently for scraps under my chair) followed by a nap on the beach. Of course I was overdressed, keeping my shirt on and all, but Dave didn't seem to mind. In Dominica, vendors came to the boat wanting to take our garbage (for a dollar a bag), or selling fruit or locally produced handmade trinkets (usually labeled "made in China"). We had to turn down offers from River Guides to venture deep into the island's terrain—there are only a few weeks left to get *Wild Hair* to Grenada before the hurricanes come. We plan to return to linger in these Caribbean jewels when we do an about-face and head home next year.

Our quick timetable didn't stop us from anchoring off the picturesque village of Trois Islets on Martinique (the hometown of Napoleon's soulmate Josephine), or keep us from buying fresh caught conch for dinner from a fellow on a boat set ablaze with letters spelling Johnny-be-Good. And the absolute weirdest thing we did (most peculiar *by far*) was take a detour into St. Lucia's Anse Cochon for a quick snorkel. Rather than finding layers of coral and fish painted brightly, we found jellyfish—lots and lots of newly hatched, nearly translucent jellyfish. Jellyfish sting, right? But these were infants, Dave and I thought, so how much zap could they pack? I can tell you, when they got stuck between our upper lip and snorkel, and also in my bikini cups, flesh went numb.

This morning, *Wild Hair* is the lone boat cruising along five miles off the Atlantic coast of St. Vincent. The day is as warm and stale as the inside of a lunch box forgotten in the sun. Sails are useless. *Wild Hair*'s motor hums at top cruising speed, but progress is slow because we are pushing, salmon-like, against a four knot ocean stream named the South Equatorial Current. This gyre came from very far away, the Cape of Good Hope actually, so today, we are navigating African waters between Caribbean Islands. The world is a remarkable place.

"I wish we could go faster," I say after a while, feeling antsy.

"I know," Dave says. "Bequia is still thirty-five miles away and we've got to get there before the customs office closes for Easter. Otherwise, we can't go ashore."

"We started at 05:30, so by my math we'll just make it. Barely." I know I'm spoiled, but the thought of being quarantined on the boat just feet from our next adventure for days on end makes me kind of grumpy.

Dave and I are skipping St. Vincent altogether because of the things other sailors have said. They say burglaries and violent assaults make it foolish to anchor alone in St. Vincent's secluded harbors. Even if we're one of a dozen boats in a bay, it's quite possible that what's unlocked on deck will disappear: outboard motors, dinghies, fuel tanks, and radios. We've done a whole heap of research online and there are oodles of stories of thieves coming on board boats, holding cruisers at knife and gun point, and demanding money and expensive gear they can use or hock. In response to the growing danger, sailors have dreamt up all sorts of ways to discourage the marauders. Some skippers turn deck lights on and stay awake all night pacing. They close hatches and rig the openings with security alarms (even though doing so turns a boat's interior into a clammy steam room). Cruisers install steel bars across their windows to prevent breaking and entering while they sleep. None of this sounds very appealing to Dave or me. So our safety plan is to steer clear of danger altogether by leap-frogging nonstop from Saint Lucia to Bequia. This makes for a long day of travel.

~~~~~

Haze pushes down from the sky, making the horizon appear fuzzy at the edges. My shirt is damp from humidity.

"We're really swimming in it today," I observe as I pick at my cotton shirt, fluff it from my back. Dave doesn't answer. He's stretched out in the cockpit with Dinghy asleep on his chest. It looks like Dave's asleep as well—his lower lip hangs loose. A small movement as innocuous as the flutter of a bird's wing in the corner of my eye catches my attention. It's a local fishing boat driving an irregular path north and south. Huh. Maybe fishermen are checking their traps, I muse. That doesn't feel right, though, because these waters are too deep. It's probably more likely someone's buying a boat, taking it for a test drive. I pick up my paperback—a spy thriller I got at the book exchange in the last marina—and leave the men to their business.

After a while, I notice the fishing boat is close—a hundred and fifty yards off from *Wild Hair*'s starboard rail. It's a local, handmade craft, blue with yellow trim and a fruity orange interior. An open design. Three West Indian men in dark rain jackets slouch with heads together. How strange. They're not fishing, like I'd expect. They're talking and shooting glances our way. Small hairs on my neck prickle. My first instinct is to turn and drive off, but *Wild Hair*'s fifty-three horsepower engine has us barely moving against this current, and I can see they have a colossal Yamaha outboard—at least two hundred horsepower. These fellows can go anywhere they want, as fast as they desire, but for some reason they're choosing to be here pacing us. Warning alarms sound in my head as my mind hunts for a reasonable explanation of their behavior.

Dave wakes up and I see him see them. I've been at the helm for three hours, so it's time to switch drivers. As we do, neither of us says a word about the fishermen. There's no need to speak. I can practically hear the clamor of sirens in Dave's head.

Ever so slowly, the fishermen narrow the distance between our boats. The instant I realize their action is intentional, that they're putting some sort of a squeeze on us, I step out of the cockpit and onto the deck. I have no idea what I'm doing but I strike a posture. I stand tall and in plain sight, arms akimbo and level my gaze directly at the three. My unspoken message: "I see you, you sons-of-bitches. I know you're there and you don't scare me. Back the hell off." My heart is racing, but I keep staring them down with the fierceness of a mama bear until (finally) their route swings wide and they give *Wild Hair* a bit of room. All at once, before I even have a chance to decide what to do next, the fishermen speed away. The local skiff grows small in the distance and vanishes in the mist.

"That was weird," I say, returning to the cockpit. "What do you think they were doing?"

"I don't know but I don't like it." After a beat he says, "Hey baby, would you grab the binoculars?" I retrieve them, along with a couple of apples, and try to settle back into my book in the cockpit. My nerves prickle.

Pretty much anyone who up and decides to go sailing (especially across international waters) wonders about pirates: what their likelihood is of meeting one someday and what they'd do if they did. Dave and I did our research and concluded there is in fact nothing romantic about pirates. Criminals—especially those in Somalia and Indonesia—do more than steal ships and booty; they take crew members hostage, demand ransoms, and extort staggering amounts of money from the shipping industry. Usually the scoundrels get away with their crimes because people responsible for policing international waters are pretty much universally confused about jurisdictions, and the laws on the books are either weak and impossible to enforce or so old they are no longer relevant.

I'd like to believe predators combing the Caribbean Sea behave differently from those in the Indian Ocean. After all, pirate lore is the stuff of tourism around these parts. The rascals of history are toasted in bars. Johnny Depp filmed a handful of scenes for *Pirates of the Caribbean* and its first sequel on St. Vincent Island. The spin is pirates are innocuous, right? Indeed, the real pirates at work in these waters are reportedly not the organized crime syndicates of Somalia and Indonesia, but small operators who are spontaneous and opportunist in their attacks. They prey on "soft targets" (those who seem particularly vulnerable or unprotected). Their judgment calls are not unlike those of lions hunting for the old or crippled water buffalo for their next meal; if the situation looks like more trouble than it's worth, they move on. Even so, attacks when they happen are not pretty. Recreational sailors like Dave and me have very recently been shot at and (in a separate incident) hacked in the face with machetes.

Not wanting to be soft targets, most sailors do a pretty good job of protecting themselves with simple discouragements. The online cruising forums sizzle with ideas. Many cruisers find it helpful to travel in convoys, for instance, or carry canisters of wasp spray (those with a twenty-foot reach make especially useful weapons). Some electrify their lifelines and keep them charged with solar panels. Others rig motion detectors to lawn sprinklers to douse anyone coming onboard. One fellow even sails around the Caribbean islands with a stash of Molotov cocktails on hand.

Dave and I get asked all the time by people back home if we carry a gun onboard. It's a fair question: boats offshore are like mini island nations, and in the absence of military assistance it's on us to self-govern. Some boaters from the United States do carry guns to stop attackers, but this self-defense strategy, by all accounts, is uniquely American. Sailors from Europe and elsewhere are of the belief that guns escalate danger because it takes only a moment's hesitation on the part of the shooter for the tables to turn, for the gun to end up in the pirate's hands. Also, they worry the guns will fire accidentally and blow holes in the hull big enough to sink a ship in minutes. Personally, Dave and I think guns are more trouble than they're worth. Most islands confiscate firearms the minute you check into customs and immigrations and return them when you depart. So, the weapons actually live inside a safe on shore most of the time.

But it occurs to me in this moment that Dave and I have no pre-arranged defenses against pirates. How unbelievably foolish. We've invested heavily in equipment to stay safe in our dealings with nature, but we haven't ever given much thought to the challenge of protecting ourselves from humans. Out here, alone, it dawns on me that we're soft targets. At once I feel as pathetic as Blanche Dubois, in Tennessee Williams' *A Streetcar Named Desire*, who said, "I have always depended on the kindness of strangers." She says that to a doctor on her way to the insane asylum.

～～～

My eyes scan the morning in a lazy arch until they land on a small craft headed straight toward us. My gut tightens.

"Holy crap, baby. They're back. This is serious. Do you want the pepper spray?"

"Without a doubt. What else you got?" In seconds the fishermen and their colorful vessel pull up within a hundred yards of *Wild Hair* and mimic our course and speed.

I head below to search for impromptu weapons. My breathing comes in bursts. Years ago I absentmindedly tossed pepper spray into a boat-bound suitcase. It was left over from my days walking through big-city parking lots after late-night meetings. I've never come close to needing to use it. I take the canister from its hiding place under the dining table and leave the envelope of cash. Next, I unlatch the wall behind the toilet and remove *Wild Hair*'s flare gun and cartridges. Then, up the stairs I go to pass the pile to Dave. Our stalkers parallel our course fifty yards away. Dread forms a lump beneath my ribs.

Below again, I pull fire extinguishers from their mounts and set them on either side of the companionway ladder. I'm thinking maybe I can temporarily blind the would-be assailants with the foam, then break their knee caps and elbows by swinging the tanks. I grab two handfuls of knives from the wooden block on the counter, put the most ferocious daggers under pillows in the salon and stateroom, and the rest in the freezer. Dinghy is asleep in the v-berth, so I shut the door so she doesn't become involved. Then I lock one of the two doors to the master stateroom aft. If necessary, I'll barricade myself inside.

I pause and breathe for a few seconds, trying desperately to think of other things I can use as weapons, or booby traps I can set. It occurs to me that what I really would love to have is a cross-bow, a couple of Tasers, a crate of stun grenades, batons, and at least three pairs of handcuffs.

A thought lifts into my awareness. "Heather, you're overreacting. There's going to be a simple explanation for this. Your worry is silly."

"Really," I say out loud to my thoughts. "*Really?*"

"This isn't what you do, Heather," my reasoning mind continues. "You're Buddhist. You're peaceful."

My flitting thoughts land on the Buddha's pirate story. When the

Buddha was alive, a violent criminal got on board a ship at sea, took away the captain's sword, and forced all the men, women, and children to line up. The pirate was going to kill them all and steal the boat. It just so happened that the captain was a student of the Buddha's. You would think he would be a pacifist, but no. The captain had secreted a dagger in his boot and he slipped it from his ankle and deftly slit the attacker's throat, killing him in a single motion.

How can violence ever be justified?

The Buddha explained it this way. By taking one life, the captain protected the pirate from the mountain of bad karma the criminal would have earned had he killed dozens. The captain acted without anger, fear, or hate toward the pirate and his swift action avoided great quantities of suffering. According to the Buddha, stopping the pirate was an expression of fierce compassion toward the pirate, the passengers, and himself.

So Buddhists are not pacifists. We act from a place of fierce compassion. When necessary we absolutely stop others from causing harm in the spirit of a strong teaching, one rooted in love and insight. I remember Thich Nhat Hanh saying on retreat that compassion is the only energy safe and useful enough to protect people, animals, and Earth because compassion is born from mindfulness, understanding, and love.

Standing in the salon below decks I ask myself, Can I act decisively from a place of fierce compassion? I honestly can't sort through tumbling emotions enough to know for sure. But I decide here and now to be as fierce as necessary to protect life. Making this commitment, I feel a little clearer in the head, slightly stronger in my body, but still completely unnerved.

~~~

Climbing into the cockpit, I discover Dave playacting. He's holding the VHF mic to his mouth without pressing the broadcast button.

"That's correct," he says. "Three men in a blue, yellow, and orange fishing vessel. They are wearing," he looks at them to assess their appearance,

"Dark blue and black jackets and traveling at three knots thirty yards from my rail." Dave is standing, hand steering the boat, and checking back and forth between our radar and the GPS. "We are at 1-3 degrees 1-7 minutes north, 6-1 degrees 0-8 minutes west, with a heading of 2-1-8 degrees."

No one could possibly be listening, and there isn't a chance anyone could come to our rescue even if they were. But the fishermen, who are tooling around without a radio, can't be sure he's faking. As far as they know Dave is speaking with the St. Vincent Coast Guard or crew on another boat lagging behind in the mist. They can't be sure. As I watch, doubt grips the trio and they start an animated conversation; they are visibly upset.

Abruptly, the men who spent the better part of an hour pursuing us rev their engine and surge their boat in front of our bow, casting a wake. They circle *Wild Hair*'s port side and disappear behind us into the vapor. The ruse worked. Dave made them believe we were in fact *not* such a soft target after all.

A wave of relief and fatigue washes over me and my limbs hang at twice their weight. I sit heavily on the cockpit cushion.

"Bastards," Dave spits. "You know they were going to beat me 'til I was unconscious, rape you, and then take whatever they wanted. You know that, right? We're *so* lucky to be alive."

"Yeah, well I was ready to nail at least one of them in the kneecap with a fire extinguisher." I smile, feeling incredibly fortunate. The two of us are frankly too stunned to say anything other than small talk and move about doing ordinary things to shake off terror. Eventually, I restow the pepper spray, flare guns, knives and fire extinguishers and open the v-berth to liberate Dinghy. When it's my turn to relieve Dave at the helm, I don't just drive but stand guard as *Wild Hair* presses as fast as she can toward Bequia.

REFLECTIONS ON
FIERCE COMPASSION

How am I to live in a suffering world? My perennial question for the Great Atlantic Teacher rolls through my head after our pirate encounter. Judging from our close call off St. Vincent, it's clear I am certainly *not* expected to let others hurt me or my loved ones or steal the things we need to stay safe. I am to *stop* people who are doing harm and not be passive. All right. But the Buddha's pirate story and my experience at sea leave me wondering how I will know for sure *when* a decisive expression of fierce compassion is called for and *what* exactly I should do.

Dave and I shared our pirate story with just about everyone we talked to in the weeks and months that followed. Listeners responded in all sorts of ways.

"Oh my! You *never* should have traveled alone," one woman said.

"Are you kidding?" said an old salt. "I've been to St. Vincent many times, anchored all over the place and never once had a problem. The people are great. I'm sure the whole thing was just your imagination."

"I would have shot the bastards if they got within a hundred yards of my boat," said a third. "No warning shots, just shoot to kill before they knew what hit 'em. But of course, I'm a retired sheriff and I've been trained in such things."

And when we tried to report the incident to an online forum responsible for keeping track of piracy on behalf of mariners worldwide, they emailed back, saying: "We are quite certain you misinterpreted the fishermen's intentions because there has never been an act of aggression that meets this description in this location."

Well. All this drives home the point that there is more than one way to respond to danger.

Still, the Buddha was clear: *stop* people from generating bad karma that will take lifetimes to clean up. Thay is clear too. He says we're not to take sides in confrontations but we are to be compassionate for all parties as we take a stand against oppression, violence, and injustice and speak out even when doing so is uncomfortable or dangerous.

Wondering how I am to be (*especially* when I come across people inflicting harm) isn't limited to sea pirates; the question pertains to all sorts of people acting dangerously and hurting others. Is there a way to stop people from inflicting harm that is more mindful, skillful, and effective than the traditional advocacy I pursued in my career?

I have to say there were a lot of years before I came across Buddhism when my advocacy was definitely *not* rooted one iota in universal compassion. When I started, I had no qualms about choosing sides, making enemies, and assigning blame. I adopted my style from others, and figured it just came with the territory. But now, I'm starting to glimpse a whole new form of advocacy—one that employs fierce compassion as it's guide and is contained completely within the boundaries of mindfulness practice. It's an approach to making change that is about stopping my own knee-jerk reactions to what's triggering me first, reflecting deeply upon the causes fueling the problem, seeing the nonviolent action that is transformative, and taking action to stop (or help stop) harm in its tracks as an expression of compassion for all involved. Is it possible for me to get so good at stopping, looking, seeing and then doing that I could be as effective as a knife blade across a pirate's neck?

Things are getting pretty scary when it comes to the harm some people are willfully causing—especially on a global scale. I'll give you an example. There was an eight-month investigation recently released by InsideClimateNews that revealed Exxon (the largest oil and gas company in the world) actually knew starting in 1977 that the burning of fossil fuel was destabilizing the global climate.[32] In 1977 Exxon knew this.

Wanting to know more, Exxon went on to fund empirical research and build climate models throughout the 1970s and 1980s. Repeatedly, *their own scientists* spent decades telling corporate decision-makers that a doubling of greenhouse gas would raise Earth's temperature by two or more degrees and the effects would be catastrophic. If you're like me, you would expect leaders of the company to pause when they knew what was at risk, look deeply into the ramifications of their actions, and act from a place of compassion. But they did not.

InsideClimateNews found Exxon responded to the science by forming the Global Climate Coalition—an assembly of industry giants who shared a financial interest in the continued use of fossil fuel. Oil companies banded together not to protect the world from danger (and earn good karma) but to cast doubt on the reality of global warming (and generate bad karma, *really* bad, like on a global scale). To raise doubt, the gang hired the consultants who, while working for the tobacco industry, were supremely effective in deceiving the public about the risks of smoking. What's more, former Exxon employees, federal staffers, and internal corporate communications all said the company had a direct hand in undermining the first global climate agreement, in preventing the US, China, and India from signing on to the Kyoto Protocol in 1998. For more than twenty years Exxon and its partners committed acts of *piracy*: they stole from people and other life forms our health, safety, freedom, and future.

Ever since the Great Atlantic Teacher brought me face to face with Caribbean pirates, I've asked myself what exactly I can do to *stop* dangerous pirates like Exxon from profiting from harm. My fiercely compassionate actions ought *not* to match violence with violence. (As Dave demonstrated, there are a variety of peaceful and effective things one can do to stop piracy in its tracks.) But the one thing I *cannot do*, if I expect to stay safe, is *nothing*. So in meditation, I contemplate the most effective ways I can protect life here and now. What's different these days is I try not to ask from a place of hate or blame, but from one of understanding and fierce compassion. The answers I arrive at are symbolic, practical, and effective acts of concrete resistance. Doing what I can where I stand, my life becomes my message. Specifically, I learn what I can about the issues; publicly shine light on the harm and suffering I witness (its causes and conditions); vote for ecologically-minded leaders; join others in demonstrations and nonviolent acts of civil disobedience; and personally cut off my financial support of pirates' products.

There is no set formula for decisive and effective action. But talking with others about ways to stop piracy is part of the solution. If enough people in society come together, speak out, and take action, then we have a chance to bring an end to these and other violent acts in order to keep this and future generations safe.

The Great Atlantic Teacher made clear that I have to act, but the teacher leaves it up to me—to us—to look deeply into the circumstances of the moment in order to know what to do.

~~~

To help me cultivate mindfulness, understanding, and love in advance of action, to prepare myself for action rooted in compassion, I wrote and practice this meditation.

Breathing in, I feel myself nourished by life.
Breathing out, I smile to those things nourishing life.
In, nourished by life.
Out, smile to life.
(Ten breaths)

Breathing in, I see the actions that are harming life.
Breathing out, I feel strong emotions arise in
    me as I bear witness to harm.
In, life being harmed.
Out, my emotional reaction.
(Ten breaths)

Breathing in, I extend compassion to myself
for the strong emotions I feel.
Breathing out, I extend compassion to others
who also experience pain.
In, compassion for my suffering.
Out, compassion for other's pain.
(Ten breaths)

Breathing in, I look past blame, deep into the root
causes of the harm I see happening.
Breathing out, I extend compassion to those perpetrating the harm.
In, looking deeply into the causes of harm.
Out, extending compassion toward harm's perpetrators.
(Ten breaths)

Breathing in, I see a healing, nonviolent
action I can take to protect life.
Breathing out, I pledge to take that step to stop harm.
In, my healing action.
Out, pledge to take action.
(Ten breaths)

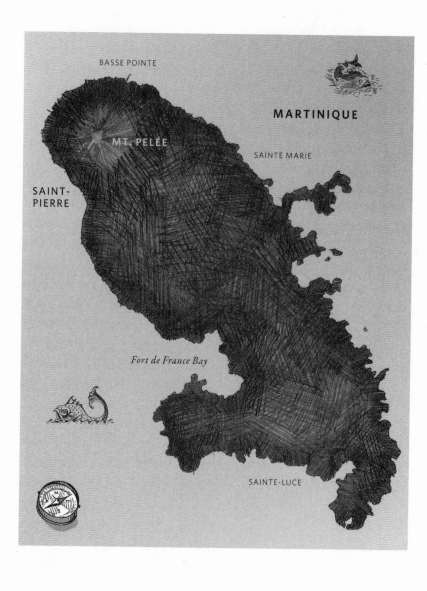

BASSE POINTE

MARTINIQUE

MT. PELÉE

SAINTE MARIE

SAINT-
PIERRE

Fort de France Bay

SAINTE-LUCE

11 THE GREAT TOGETHERNESS

Beyond *Wild Hair*'s port rail, the wrinkled emerald mass of Mt. Pelée stands more than four thousand feet above sea level. That's two thousand feet taller than Indonesia's Krakatoa and a little bit higher than Italy's Mt. Vesuvius. As mountains go, it looks pretty unremarkable. It's a dome of rock tall enough to snag clouds on the irregular top. Giant ferns give a fuzzy-green cast to the highest, steepest slopes. Gashes radiate from the peak and the divots hint at the existence of dozens of rivers. Halfway down, where trees and crops frost the earth, the land becomes broad and slanting. I can't see that anyone lives in the hills. But at the water's edge, the city of Saint-Pierre—population four thousand—hugs the crescent-shaped harbor. Fishing boats in botanical Caribbean colors, red-roofed buildings, and arching palm trees dot the shore. A blue sky and glowing blue ocean frame the whole.

Wild Hair floats fifty yards from the beach, above the remains of a couple dozen ships resting on the ocean floor. On this languid morning, it's hard to imagine the hell that opened on May 8, 1902, when Mt. Pelée exploded and sent a wave of volcanic gas down the hillside to incinerate the treasure known as the "Paris of the French West Indies," along with the ships in her port. Thirty thousand people died where they stood, their bodies crisped in identical postures: arms lifted and hands shielding faces from flying cinders and heat.

I'm reading *The Last Days of St. Pierre: The Volcanic Disaster That Claimed 30,000 Lives* [33] and author Ernest Zebrowski, Jr. reports that the remains of people in Pompeii were found similarly frozen, with hands up. The book's historic photos of the aftermath show there was nothing left of the place but scorched earth and a suggestion of architecture in the surviving masonry, a ghostly honeycomb grid pattern. This Sainte-Pierre I'm seeing now sprouted from the ashes.

Wild Hair isn't going anywhere. Chris Parker says the sea around Martinique is uncommonly violent and utterly impassable. It's hard to imagine—from where I sit, the world is tranquil. But he's adamant. Distant storms are happening in both the north and east, he says, and they are radiating massive swells powerful enough to eat us up and spit us out. Our voyage is taking us north now, and again I feel the pressure of having to be outside the hurricane zone by a certain date. (This time, we're working our way to the state of Georgia.) I understand lingering in Martinique isn't what anyone would call a hardship, but Dave and I will probably have to make up the week's delay by missing out on a place we've never been to. The thought of missing out makes me want to get going.

"C'mon, Great Atlantic, don't you know I've got an agenda?" I direct my outburst into the day. "I've got things to do, places to be." I drum my fingernails in rapid succession on the teak table in the cockpit. Sunlight flickers across the water. My inner voice replies, "Heather, do you really think the Atlantic gives a flying fig about your big plans? Nature's in charge here. It'll decide what you will and won't do."

"All right, all right." I take a sip of hot coffee, turn toward shore, and try to let go of my ambition. A week, I think. I guess this is my chance to dig deep into this sleepy village, see what it can tell me about how to be in a suffering world.

~~~

Dave, Dinghy, and I have been months on the move. The three of us spent last summer in Wisconsin, with *Wild Hair* in the boatyard in Grenada (just eighty miles from the coast of Venezuela). Boat repairs and maintenance consumed the month of November, but we snuck in a few parties and tours with other cruising couples from Jamaica, England, South Africa, and the US. On Thanksgiving, Dave and I celebrated at the local beach bar with seventy American sailors (the holiday felt down-right authentic, with the Green Bay Packers-Detroit Lions football game on the big screen). For a really wacky outing, Dave and I hurled our bodies into a Grenada "hash"—a four-mile trek up and down some steep and soggy terrain, across shallow rivers, and through muddy valleys. It was a slog that basted us in inches of muck, but the camaraderie between both cruisers and locals was warm, and the swim in the quiet brook followed by a meal at the rum bar revived us.

Dinghy turned into a mud puppy on Grenada. The instant she saw her first land crab disappear into a hole near the boatyard, she stuck her arm through the opening all the way to the shoulder. The feat turned her pure white fur into a grubby mess, but she could hardly care. Now, she roots around in every hole we pass.

We left Grenada at November's end and hiked and snorkeled our way through Carriacou, the Tobago Cays, Mustique, Bequia, St. Lucia, and now Martinique. I must be addicted to roaming. Now that we're stuck, my muscles lunge like a horse in a gate ready to race. I close my eyes and take a breath.

My mind wanders back to a strange episode that happened at our house in Madison last summer. I wanted to grow vegetables (it seemed like an especially good thing to do to celebrate being on land), so I was outside stripping grass from the only patch of sunlight in my wooded lawn (actually missing *Wild Hair* and sailing something awful) when, after several hours of hard labor, I sat back on my heels and noticed a robin hopping eagerly just beyond my reach. It occurred to me he had been trying to get my attention for some time.

"Oh hello you! Are you begging for a worm?" I asked. I hadn't realized a robin could be as smart as a dog. "Well, I hate to tell you but there's
no worms here. There's too much clay." (This fact didn't bode well for my
future veggies.) The robin continued to hop, so I made conversation and
said, "Hey, do you know my friend the dolphin?"

"Yes and no," he responded. "She is my cousin but we have not met."

What? I startle at the answer. Where the hell did that reply come
from? His eyes stayed on mine. The bird's head tilted and then he hopped
again. All at once, it dawned on me that the Wisconsin songster was a
teacher of wilderness ways, as divine and knowing as the Great Atlantic.
The bird in my yard was another particular jewel set in the continuous
web of life, whose polished facets perfectly reflected the universal wisdom
contained in the whole of the wilderness. Suddenly, I wasn't separated
from the Great Atlantic Teacher. My two lives (inland and at sea) were
instantly one and the same, varying only by degree, like water and ice.

The little critter helped me see wilderness in a new light: not as something far away or unique, but rather as the ordinary state of being for all
things. Civilization may *appear* to be something other than wilderness,
because it distances me from untamed things. I may think my city and
the farmland that surrounds it bears no resemblance to the ocean wilderness or jungle or mountaintop. But wilderness (governed by the laws of
natural order) persists around and through the full range of Earth's ecosystems and all of humanity's schemes. The robin told me it didn't matter
where I lived. He got me noticing the myriad ways wildness pulses in the
web of life on my streets, in my house, through my work, and in my body
and mind.

I started seeing wilderness everywhere and at every scale. I saw bugs
as wild when they appeared in my home. Wilderness was in the blue
veins of cheese in my refrigerator, in the blossoming patches of mold in
my shower, in the cracks of my driveway, and in my aging flesh. Dinghy,
I realized, was undiluted wilderness in my home. Color, light, stars,
sepia tones in old photographs, and chapped lips were suddenly proof

that I was onto something organic and beyond my control. Eventually I comprehended wilderness is everything that happens that I don't plan. Wilderness is what I couldn't possibly create with my limited know-how: new life, the spontaneous experience of love or embarrassment, flowers, clouds, rusty tools, a puddle of water, and hills.

As I sat on my cushion with this growing awareness, I eventually came to see meditation as my tool for witnessing and understanding what's happening in the wilderness. I realized that my habit of reasoning things out was ultimately a failed attempt to contain and control the wild world. I can't reason my way toward enlightenment; it can arrive in the form of a robin. So meditation is my remedy to reasoning; it's a skillful way of tuning in and jiving with the ubiquitous wildness of my life.

Dinghy jumps onto the cockpit table and sniffs the fresh-baked pumpkin bread on my plate. I pinch a corner off and put it next to her front paws. She eats it greedily. "Good morning, wild thing," I say. Dave climbs the ladder from below and stops suddenly.

"Is that boat sinking?" he says. I turn to look. A short distance behind me is a catamaran. The mast tilts, the starboard hull sits low. A white and pink skiff circles the vessel.

"Yup, by golly, it's taking on water."

"Wanna help save it?" Dave's eyes sparkle. Suddenly, I smell adventure.

~~~~

"We're here to help," I say, as our dinghy kisses the hull of the thirty-eight foot Fontaine Pajot. I hand the dinghy painter to the young man greeting us, expecting him to tie the line to his vessel, but he lets go of the rope and reaches for the pump in my lap. A half-dozen people scurry about on deck and the boat is a cacophony of French conversation.

"That's my pump," Dave says firmly. "I'm not giving it to you. I'm coming aboard." The young man stands. I pull our dinghy line from the sea and hand the soggy mess to him a second time. The fellow takes it but doesn't pull the dinghy in tight or secure our connection. The looseness makes our inflatable bounce out of sync with the large boat, and it's a struggle to climb aboard. Clearly our greeter isn't a mariner. Judging by the hubbub and confusion onboard, I guess the others are equally in the dark.

Thirty feet of hose looped around Dave's neck make him as gangly as an octopus as he steps on deck. He carries a three-foot handle and I'm gripping the cast-iron pump. This is the first time we've used *Wild Hair*'s emergency manual pump, but I know each pull of the lever is supposed to draw a gallon of seawater from the bilge and put it overboard. Dave steps below deck to assess the problem. His legs disappear in thigh-high water.

A woman and two men at the center of activity are upset and speaking French very quickly. Four others onboard—including the greeter—mill around. Every once in a while a bucket of water passes hands and spills overboard. No one says hello or gives Dave or me instructions, so we get busy attaching the long hose to the pump's intake and the short hose to the outflow. The idea is to snake a continuous tube from the lowest point in the catamaran through the cast iron contraption and up and out to the sea so we can suck the water overboard. We finish laying the hose and I'm sliding the pump's handle into the cast iron slot when a great-looking, way-too-young-for-me French fellow comes to my side to operate the mechanism.

I want to show him how it works, so I step onto the board the pump is mounted to and lever the handle forward and back. After a few pulls, gobs of water tumble overboard. He smiles and takes over.

"Thank you for your help," the young woman says.

"Is this your boat?"

"Yes. We live on shore, but my husband called this morning from Paree to tell me it was sinking. A friend noticed za boat was tilting and called him, and then he called me. I phoned everyone I knew who maybe could help." It registers that the older people sitting in the cockpit corner look like her parents. The woman introduces herself as Melonie.

"Do you sail?"

"No. It's my husband's boat. These are his friends." She points at two of the Frenchmen: one is manning the pump, the other is on the phone.

"Did you stop the leak?"

"Yes." She nods at the man on the phone. "Pheeleep swam and plugged all za through hulls. Maybe za toilet took in water, I don't know. But za sinking stopped."

"Uh-oh!" The man at the pump says. He holds the handle up; the bottom three inches snapped off, fractured from the stress of levering. I remove the stump from the slot and he reinserts the handle. The rod is shorter, but the pump still works.

"Let's get some buckets going." Dave says to the crowd. Two buckets materialize and a line forms. I'm on the end, responsible for dumping water overboard, but the boat is littered with equipment and the wide, inert frames of Melonie's parents, who block the only reasonable access point. They remind me of a pair of giant mushrooms growing in the shade. I do my best to work around them.

"*Uph.*" I look and the pump volunteer is making a small pile of three-inch aluminum tube sections. He's kneeling on the board, now; the handle shrunk by half. Sweat drips from his nose.

"Dave," I call below, "I'm going to *Wild Hair* to get another handle." I take off in the dinghy and return in minutes with a hefty sixteen-inch stainless steel rod. I hand it to the kneeling, exhausted Frenchman. It's a short stick for the job but at least it won't crack.

There are fewer people onboard now. The spectators must have left while I was gone. Phillip is still on the phone. Actually, he's doing three things at once: talking into the phone wedged into his shoulder, scooping buckets and handing them to Dave, and taking deep drags on a cigarette. This guy is as hyper as a Wall Street trader.

The fellow manning the pump slumps on the ground. "Fini!" he says panting. Melonie goes below to check.

"OK, *oui*," she says. "Za water, it's below za floor."

"We will leave you to clean up from here," Dave says, coiling hose.

"*Merci beaucoup.*"

"*The pump est tres bon!*" Phillip says, snapping a lighter to the fresh cigarette in his lips. He returns to his phone conversation.

"*Magnifique,*" says Melonie.

The man on the floor rolls off the mount without comment. Dave and I move to help him stand but he waves us off.

"It's good to know the pump works," says Dave. "We just need a long handle that doesn't break, I guess. Thank *you* for teaching us something about our equipment!"

~~~~~

With the dinghy nestled between other inflatables at the public pier, Dave and I lug two forty-pound plastic bags of laundry on our backs down the streets of Saint-Pierre. The sidewalks are paved but marred with holes deep enough to snap ankles. Cars the size of large picnic baskets speed by a little too close for comfort. Around us, graceful French architecture snuggles with charmless concrete buildings in a hodgepodge of styles. Wrought iron balconies sag from the weight of unruly container plants. Beach umbrellas advertising sodas, beers, and mineral waters sprout from café tables. The exteriors of buildings glow yellow, orange, green, brown, and blue. Patches of exposed brick and layers of Caribbean mold give the whole an authentic French shabby chic décor.

Gaining elevation, we move north from Rue du General De Gaulle beneath neon signs illuminating words I cannot pronounce. We pass a bakery, shoe store, pharmacy, and a smattering of banks and bars. Sidewalk menu boards advertise "*Les pâtes de Fruits de Mer.*" Inside, business women in colorful, tightly sculpted dresses and four-inch heels sit at oceanfront tables, linen napkins draped across their laps. We cross the street to follow the pedestrian path. Old women rhythmically sweep tiled front stoops. *Shish-shish-shish.* Their eyes are down, lips pursed in concentration. Powerful smells of buttery pastries, fresh fish, and yesterday's wine tantalize us from all directions. French ballads, ringing phones, television programs, and women's laughter spill like a soaking rain from behind the laced curtains and hurricane shutters of second-story apartments. This is life in modern Saint-Pierre.

Dave and I step inside the laundromat and drop our loads. I roll my shoulders in an attempt to loosen them. The small room is empty, but five commercial washing machines and a pair of enormous dryers, each capable of heating and tossing three wash loads at once, are ready to accept our dirty clothes. With fistfuls of Euros in our pockets, we will make short work of the big job. I'm ecstatic.

"Do you want to hike Mt. Pelée?" I ask Dave.

"That sounds fun. I could use some exercise."

"Our cruising guide says there's a trail, but it doesn't say where or how to get there."

"We can figure it out. We'll just keep going up!"

Buying a baguette, brie, and bottled water at the 8 à Huite supermarket the next day, I ask the clerk where to find a bus station. She signals a route and we walk north a few blocks to a dark green repurposed gas station. A dozen well-dressed commuters hang around. I approach a man wearing a light blue short-sleeved shirt who seems to be in charge.

"*Pardon moi. Parlez-vous anglais?*" I ask.

"A little," he says, while his hands arrange a wad of tickets.

"We want to hike . . . uh . . . promenade Mt. Pelée. What *autobus?*" I lift my hands and make my face into a question.

French words zing from his lips. Then, the man looks up with a start and rushes Dave and me sideways. A large bus rolls onto the pavement where we stood. I retreat to the concrete platform that once held gas pumps.

"This bus?" I point.

"*Non.*" He gestures for us to wait. Dave and I stand on our island of safety for more than ten minutes as people, luggage, taxis, and buses come and go. Sleek new buses glide into and out of the station, whisper quiet. Wide, panoramic windows, air conditioning, and plush reclining seats look downright inviting. No one says much; everyone seems to understand the goings on but us. Finally, the attendant signals to us.

"Here." More French cascades from the man's mouth as he holds out his hand. Dave pulls a fistful of bills and coins from his pocket and our helper sifts through the stash, taking what he needs before motioning us to board. I've learned to trust the community while traveling.

"*Bonjour. Parlez-vous anglais?*" I ask the woman driver.

"*Non!*" she says definitively.

I am stumped by what to do next. The driver cranes her neck to look past me to the ticket man who is speaking to her. She nods. The doors shut and the bus accelerates. Dave and I are off, but who knows where to?

The bus carves wide turns down narrow cobblestone lanes. We cruise smoothly across a bridge, and the commercial center gives way to a neighborhood where steel bars protect windows and graffiti marks walls. The smell of delicious cooking penetrates the bus's climate controlled air. Older people amble the sidewalks carrying canes and shopping bags. Young people drift in packs. Soon we quit the city altogether and we are on a fast highway, twisting and gaining elevation.

The bus veers right and stops abruptly. An elderly woman heaves herself aboard, one step at a time. Waves of French pass between the new passenger and the driver. The woman sits in the front row and the conversation continues as we motor. After a while, the bus swerves to another stop. This time, the fellow sitting near the back of the bus gets up to leave. "*Bon après-midi,*" he says, stepping out.

"*Au revoir,*" says the driver. Things continue like this as we wind our way—past picket fences, churches, community centers, and country homes. Occasionally the road turns and my window frames a magnificent view of the city below and the broad Caribbean Sea. It occurs to me the bus is following Mt. Pelée's lava path upstream.

After about an hour, the driver swerves right and calls, "*Ello.*" One eye, reflected in her rear view mirror is trained on me. Her hand motions us forward. She unfolds an accordion map, puts a finger on our location.

A line noodles to the left marked D39. Looking serious and speaking helicopter-blade fast, the woman points out the window to a road going left and up.

"We go there?" I ask. She nods, her hand sweeping in a half-circle to reinforce the message.

"How do we get back?" Dave whispers to me.

"*Où est the autobus to Saint-Pierre*?" I ask in appalling pidgin French as I gesture downhill. Again, the barrage of words comes quickly as she points to the map and waves her hands. After a few minutes of charades, I come to a crossroads between sort of understanding and knowing further clarification is impossible. I give up asking.

"*Merci beaucoup*," I say.

The woman shrugs. To her, the kindness is small.

"*Non, non, merci!*" I insist. "*You are magnifique!*" The woman laughs.

We exit and walk uphill into the French countryside. A chill penetrates the air. Trees and nearer hills obscure the mountaintop. No one is in sight. Three milk-chocolate-brown cows chew their cud.

"Hello, cows. Do you want to come with us back to Wisconsin?" I ask.

"Look at that one over there, under the tree," Dave says. I look. "She thinks you're nuts."

"Oh, does she now?" The cow flicks her tail as her teeth gnash. "That's just because she hasn't been to Wisconsin."

"Are you going to buy her a wool coat?"

"I guess a Caribbean cow probably doesn't want to move north, does she?"

The road bends and suddenly a 360-degree view opens. We are on Martinique's highest ridgeline. East is the deeply hued Atlantic, west is the sparkling turquoise Caribbean. South are the triangular points of geologically older volcanoes—the Pitons du Corbet—and north is the broken rim of Mt. Pelée.

"This is incredible," says Dave. The vista is spellbinding. We stand in silence for a long time, breathing, taking it in. The Atlantic coastal city of Le Lorrain carpets the lowland 2700 feet below.

The way I understand it, the eruption of 1902 started spitting and coughing almost three months before it actually detonated, giving people plenty of advanced notice to clear out. There were plumes of steam. The ground trembled and terrified livestock and dogs. Snakes and small mammals living in these parts dashed downhill and into the city and soon Saint-Pierre overflowed with pests. Foul sulfur smells were in the wind and the telegraph connection to neighboring Guadeloupe severed. While all this was happening, the residents wavered between fear and calm. They kept telling themselves everything would be all right, because they remembered Mt. Pelée's 1851 blow that did nothing but dust fields and trees with ash before it quieted.

But the mountain didn't fall silent this time. By early May, people were terrified—the ground emitted a constant growl and the Earth often shook. A black and expanding cloud (one that churned with lightning) turned day into night and the streetlights were necessary at noon. The telegraph cable pointing south to Dominica went dead. Then the volcano did something no one in the west had witnessed: it sent a flood of boiling muck (a *lahar*, according to author Zebrowski in *The Last Days of St Pierre*, more than 500 feet wide that traveled 60 miles-per-hour) down its creases. The blistering deluge killed four hundred people in the nearby village of Le Prêcheur.

Why didn't people leave town? Well, a handful *did* leave Saint-Pierre (women and children mostly), but the refugees arriving from the countryside more than made up for their numbers. Those who stayed assumed that lava, if it spilled, would be slow and follow the path of the riverbeds. Also, they worried about too much ash collapsing homes, fires from cinders, thieves running rampant, the trek up and over mountain roads, and the lack of food and a place to stay when they arrived to a new locale.

Standing near the top of the world on this sublime day, the mountain apocalypse doesn't seem real. My eyes scan the slope closer in and stop when I see my new favorite Caribbean vegetable: christophene. The citron green vegetables (some call them chayote) hang like pears under wide, spruce-colored leaves.

I also notice another of my favorites: dasheen. Sets of four and five leaf stalks spring vase-like from evenly spaced taro roots. The ten-inch heart-shaped leaves are in various stages of unfurling and dipping to the sun.

In a quarter mile (and another two hundred feet of elevation), the road turns and the summit comes into view again. Mt. Pelée's top seems no closer.

"Baby, I'm tired," I confess. "We've been walking uphill for hours."

"I'm not surprised. You were wiped out with the flu a few days ago."

"But I'm sick of being a slug." I take a breath. "How much longer is it to the top? I really want to walk the edge of the caldera."

"Don't worry about it. Look how high we are. The view is so pretty, but the clouds are coming and we won't see the ocean if we go much higher. Besides, I think we've climbed 1500 feet above where the bus let us out. This was a nice walk."

"I love you."

"Check out the top. See how the blast exploded the rock." Dave turns me around and wraps me in his arms. Before us is the vent of the eruption. It's exposed because an immense section of the rim broke away. Mt. Pelée is named for the shape of a man's head. But the dome is sunken now.

"Mt. Pelée's bald head was chopped with an ax and its brains spilled out," I observe.

"Hmm. Gruesome but true. How 'bout we eat lunch at that restaurant we passed, the one with the courtyard garden."

"I really love you!"

The restaurant could feed busloads, but we are its only patrons now. We pick the front table overlooking Mt Pelée's hollowed cranium and order the special: a three-course African-French fusion meal. The waitress puts small plates of blood sausage on the checkered tablecloth.

Cooked pig's blood and milk congealed is pretty rough for an aspiring vegan, but I tell myself it's French and ignore the gelatinous texture, swallowing before I lose courage. The main course is a serving of creole mahi-mahi alongside jerk chicken thighs, surrounded by red rice with beans, fried plantains, and slaw. The hot and spicy meal is yummy, but I can't shake the feeling that Mt. Pelée is our lunch guest. Although it's on its best behavior at the moment (it's not expected to blow for another three hundred years), the caldera isn't company I trust.

Martinique elected its first black senator to the French national government just months before Mt. Pelée went off. The powers that be went nuts at the idea of a black man in power, and they completely wigged out at the thought of evacuation because the *next* election was only days away and *another* black candidate was gaining popularity. If those with means fled the city and didn't vote, people without means would stay behind and put a second black candidate into office. So, the mayor (who was also a businessman) plastered the city with posters characterizing fear over the volcano as hysteria. Martinique's governor was likely in cahoots too—he threatened public employees looking to evacuate with termination and a stain on their record. Even the publisher of the local newspaper campaigned aggressively for more affluent people in the countryside to relocate to the city, because he knew they'd vote for the white candidate. The collusion of politicians, businessmen and the media (over many months) is what historians say convinced residents to attempt to ride out the volcano in the face of what must have looked like the end of the world.

The blind political determination of civic leaders (well beyond the point of reason) got themselves and 30,000 others killed. How, I wonder, did they think themselves so separate from the people they served and so impervious to their environment? The depths of their hubris are difficult for me to fathom, given the dramatic introduction I had to my interrelationship with the wilderness a little more than a year before I started sailing on *Wild Hair*.

In June of 2006, I found myself walking under the canopy of my umbrella at the edge of a vineyard. It was the second of a three-week retreat at Thay's Plum Village Monastery in Bordeaux, France. The soft rain fell straight down, so it was possible for me to keep my feet and the hem of my skirt completely dry under the tiny umbrella. Thay had just offered a teaching on the reality of our interconnectedness with the cosmos (the great togetherness he calls "interbeing") to a crowd of practitioners, and his words echoed in my head.

"Everything is a continuation," he said. "Because everything is a continuation, nothing exists separate from the whole. The idea of a separate self we call this body is an illusion."

I went on retreat this time not only to study and practice mindfulness but to become ordained into Thay's core community of practice, the Order of Interbeing. The way I understood it, the ordination was equivalent to that of a chaplaincy in other traditions: ordination meant I was committing myself to a career of helping others discover the teachings and supporting the community of practitioners. I had been meditating on the idea of interbeing for years, but I still found the concept challenging.

"I kind of get that things evolve from one moment to the next and that time is continuous and connected," I muttered to the ground, "but I don't understand how that relates to space, why that means nothing contains a discrete and separate self." The warm rain soaked into the mowed grass as I stepped. "Literally, how is my umbrella a continuation of my shoe? How is my right hand related to that grape leaf?"

I took a breath and gave up grasping at thoughts for the moment. I decided to simply walk and drink in the beauty of the French countryside. I remembered the teachings themselves were supposed to be like rain; they needed only to land and soak into the fertile soil of the subconscious in order to nourish the roots of understanding. Insight, the monastics assured me, would eventually bloom. But how long was insight going to take?

I was on my way from the Upper Hamlet, where Thay and the monks lived, to the Lower Hamlet where I was staying with the nuns. I pictured myself stepping through dharma rain as I casually admired the gnarled branches of the grape vines planted in impeccably neat rows. I saw the strong wire fencing holding branches toward the sun. I wondered at the age of these mature and fruitful plants.

All at once, my vision shifted and everything was different. Nothing changed in the physical world; the muted light, green vegetation, and black asphalt were as they were a moment ago. Raindrops continued to patter the umbrella. But a titanic change occurred nonetheless. I stopped in my tracks and looked with startled attention.

It was as if what my eyes saw was interpreted by my mind as a reverse image. I'd had a similar experience looking at playful optical illusion pictures: Edgar Rubin's vase in the center of a piece of paper turned into two faces in profile; the classic young miss with head turned away became an old woman with a downcast gaze; Joseph Jastrow's sketch of a rabbit changed into a duck. But on that day, my vision shifted into what I could only describe as a lively, gelatinous state of being. The explanation is going to get tricky here, so stay with me.

The most classic teaching describing interbeing (the reality that nothing is truly separate from the whole of everything else) is the metaphor of the ocean and the wave. A wave, trundling merrily along, may think it has a separate self-identity (after all, it has a beginning, middle, and end, and it is taller, shorter, faster or louder than others of its kind), but it is made of water. All waves are made of water; the same water. The waves inter-are, and when we look past the separate identity of a wave we see the ocean and it is deep and stable, rich and powerful. On that day, when I had a vision of the world as some sort of electrified goo (a type of matter so unique, omnipresent, and free-flowing that it didn't even bother to discriminate between substance and air as it quivered),

I caught a glimpse of the reality of our connection. I saw the oneness of the cosmos as a single organism buzzing with joy.

I get it, I thought. I'm looking at the visual reality of the Buddha's teaching on interbeing. And the quivering vibration animating the whole, connecting us all, it's . . . it's . . . *love*. (I didn't think of this word. It leapt into a gap in my consciousness.) I'm *seeing* love, it's a substance and it's *everywhere*. It's absolute and inviolate. Nothing exists outside of love, so it cannot be harmed. This was really something.

The next thought that entered my mind was loaded with craving: can I hold onto this perspective forever? Immediately, I recognized the clinging as a mind trap and amended my question to a statement: I wonder how long I will have this altered perspective. Not wanting to disrupt my ecstasy, I took a careful step down the hill. The vision stayed stable, so I took another step. As I walked, the goo responded to my motion with acceptance and delight. I made my way down the road, across the bridge, around the bend, and through the charming stone buildings of the hamlet. The perspective held fast even as I greeted my retreat friends. I noticed people possessed a signature hum, a unique energy pattern that of course was soundless as it vibrated. Once, someone approached me from behind and I sensed their signature energy as distinctive as a fingerprint before I saw them. When I turned to see who it was, I exclaimed, "Oh—of course it's you. You are lovely today!"

~~~

"Isn't it weird to sit near Mt. Pelée's top—the gateway to hell—enjoying a civilized meal?" Dave asks, licking jerk sauce from his knuckle. He looks toward the summit. The waitress delivers two dishes of coconut bread pudding with bourbon sauce. The smell of caramel beckons me to eat more even though I'm stuffed.

"I can't even imagine how crazy it must have been," he adds. "Hey—tomorrow—let's wander into town and see if we can locate any ruins."

"Unless Chris Parker says differently, I'm in," I say. "Let's do it!"

The next day, a loud speaker blares as a crowd cheers. Noisemakers clang. We're crossing the town's main drag when six bicyclists traveling in a tight pack shoot past at thirty miles an hour. A car is among their ranks and it sounds a rising alarm. When there is a break, Dave and I dash across the road to join the crowd lining the far sidewalk.

"What's happening?" I ask two men dressed in suits.

"Eh . . . *course cycliste*," the first fellow says from around the cigarette between his lips.

He lifts his wrist, fingers pointed down, and wags his hand toward the street. I look up the road at the cyclists' backs, plastered with numbers.

"Ah," I say, nodding and smiling.

The second man points to the street sign and Dave reads. "Rue Victor Hugo. Ah! Victor Hugo Course Cycliste."

"*Oui, oui!*" The two men exclaim, happy to communicate.

More riders speed past. Rubber tires make sticky-peeling sounds as bikers zoom by. The wind they generate billows our shirts and kicks bits of grit into my contacts. Cyclists navigate parked cars, pot-marked asphalt, and heaving cobblestone.

A car driving downhill on a cross street tries to turn left, causing chaos on the course. Spectators holler and bang the hood for the driver to stop. Four riders almost collide but recover.

"This is insane—they haven't even shut down the street!" Dave says.

"Holy crap—" I say.

"One distracted driver—like us in a rental car—could take out a dozen people."

～～～

At the far end of town, a thirty-foot bluff divides the main road from the waterfront. The volcano museum sits on top of the bluff. Beneath it lies the ocean, a narrow beach, a car park, and a string of ruins. These are the remnants of warehouses—twenty-foot-wide, two-story structures that shared common sidewalls. Warehouse Row was the center of Martinique's shipping district. But today, a mini forest of weeds and ferns sprout on the knife-edge tops of wall fragments. Palm trees fan lazily in the breeze. One nearly intact building stands to offer a sense of what was, but everything else looks chewed up to a fraction of its original form. Bits of wall containing a doorway or a window remind me that this was human habitat; people watched the sunset through these frames. Farther up, walls disappear altogether, leaving only a ghostly suggestion of architecture in patterns on the floors. Unblemished tiles linger in the open air as if waiting for new tables, chairs, sacks of coffee and cocoa, or kegs of rum. The whole of it feels tragic and sad.

Dave and I enter the one-room museum. "*Bonjour*. How are you today?" a woman greets us. She is movie-star beautiful, a tall, dark goddess, lightly perfumed, with perfect hair and high heels.

"Good morning. You speak English," Dave responds.

"*Oui*. I am here to tell you about ze museum—a gift from American volcanologist Frank Perret."

"Oh, good."

"In 1929, when Montagne Pelée erupted, Monsieur Perret examined za *montagne* and determined it was safe. He rightly told the people of Saint-Pierre they need not fear."

"Wait, what? There was an eruption in 1929?" I ask.

"*Oui*. But it was miniscule compared to 1902. Saint-Pierre's friend, Monsieur Perret, built this *musée* as a present to the people. You will discover on this wall photos and stories of Saint-Pierre before the destruction. On ze other side is . . . eh . . . the depiction of post eruption."

"Tell us about that—" Dave gestures toward a bell in the center of the room. Its two-foot diameter is melted and falling apart.

"Zat is the bell from the cathedral. It did not do too well!" Her smile radiates. "Please, enjoy!" The woman gestures toward the room. "And please, before you depart, consider putting a donation to our *musée* in ze wooden box at the door. *Regardez autour. Merci*." Her attention shifts to the next couple entering. "*Bonjour*!" she begins—this time continuing in French.

The city preexplosion was seven times larger than it is now and quite sophisticated. Displays and interpretive signs reveal mansions, streetcars, public schools, and electric lights. The laundromat where we washed our clothes was once a Parisian boutique in the heart of the shopping district. In panoramic images, gentlemen walk about in three-piece suits as horses pull trollies down tracks embedded in cobblestone. Corseted ladies promenade in pairs under parasols. Pictures also showed the beach strewn with carts, barrels, stacked crates, rowboats, and people. In one shot, a half-dozen three-mast ships sit at anchor. I cross to the other side of the room and find melted porcelain plates, spoons, and water glasses stacked and fused. It's all pretty horrific, so I'm happy when we exit and head across the street to the ruins of the old theatre.

The grand flight of stairs rises from the street to the floor of a mezzanine lobby, but all the auditorium's walls are blown away. We climb a staircase that wraps the remains of a three-tiered fountain and discover what's left of the cultural hub: a sloping hole with a platform at one end. Faint echoes of Molière's masterwork *Tartuffe,* followed by cascades of laughter seem to mix with the sounds of surf. In my head, musicians and a conductor fill what looks like an orchestra pit; they accompany French opera star Simone Logien in her captivating portrayal of Salomé. Stone remnants contain morsels of magic. A sense of artistry somehow hangs on.

Sitting in the middle of the ruin, like an audience member, I try to imagine what the moment of detonation was like. The way the museum described it, a wave of steam, gas, and cinders ripped sideways from the mountaintop. It wasn't lava that destroyed Saint-Pierre, but a pyroclastic surge—black and heavy—that poured down the slope, reaching the city in less than two minutes and shooting people into the air, cooking lungs, demolishing these stone walls, and igniting and sinking wooden ships. The rum warehouses behind me detonated. In response to the blast, a tidal wave submerged the coast. On its heels, a second explosion blew seven miles into the atmosphere. The mushroom cloud billowed fifty miles wide. Fire leapt from Mt. Pelée's torn flank, and only then did lava cascade. According to the crew of a ship sailing on the north side of the island, sudden fire erupted on the land touched by the pyroclastic surge, and it drew a tremendous wind. The gust was violent, coming at the same time as a deluge of muddy water and ashes as the volcanic steam quickly condensed into rain.

Map in hand, Dave and I wander the town. Wherever I look, ruins appear. At one corner, someone plastered a sign on a singed, two-story stone husk: *á vendre* (for sale). Down the block, a modern, mirror-glass office building sits behind a four-foot-high jagged stone fence—what's left of an earlier edifice. The new proprietor mounted a mailbox into the ruin. On a residential lane, one- story façades of former row houses have contemporary address numbers, and the preeruption doorways form chic entrances to private courtyards of upscale homes. The Saint-Pierre of today is built in a cemetery. It is a way to honor your ancestors, I suppose—to keep them alive in your heart.

～～～

I spent this morning—our last day in Martinique—scuba diving the most colorful reefs I've ever seen. This afternoon, Dave and I bought fresh fruits, vegetables, and baked goods and readied the boat for tomorrow's departure. We took time to catch up with family members by phone. In a few minutes, the two of us will go out for dinner at a pretty waterfront restaurant. But presently, I sit alone on *Wild Hair*'s transom as the sun sinks into the sea. Ripples slosh against the hull, and *Wild Hair* shifts subtly. The catamaran that almost sunk floats upright in its place behind us. I decide to play a made-up interbeing game by pointing my finger at "myself." First, I lift my arm and point to the sky and try to imagine the sky as me. I am air, dear sky. Oxygen is in every cell in my body. I look past Earth's atmosphere into the coming darkness. I am stardust, too. A phrase of Carl Sagan's comes to mind, and I say his phrase aloud, "I am starstuff pondering the stars."[34] Stardust, you and I are companions. This is fun.

I close my eyes and breathe some more. After a few breaths, I lift my finger again, this time only as high as *Wild Hair*'s deck. Ah, boat; I am you. You are my strength and motion. I am your consciousness. When we go, we go together. When we stay, we stay as one. We inter-are. I close my eyes and smile, thinking that one was *way* too easy.

My finger lifts toward Saint-Pierre and I think: I am the ticket man at the bus station managing people and resources and I am the bus driver helping strangers, making friends. I am the blood of French pigs. I am the cyclist sweating and pumping my legs, feeling strong, and I am the young wife struggling to keep a boat from sinking. I am a museum docent— knowledgeable, gracious, and desirable. I am a garden wall: singed and crumbling, and still beautiful in my entropy. I am the old political leaders of the town: confused and in need of wise friends. Thank you Martinique for being my community.

My hand rises again and I point to the mountain. Oh—I am Mt. Pelée and Mt. Pelée is me. We are one. You and I have the same core—a dynamic center that runs hot and cool. Our bodies support numerous life forms. We are dangerous because each of us has erupted in our past and hurt others; we will likely do so again. But in this moment, we are peace; we allow ourselves to be shaped by one ocean. Your quiet presence is a lesson in the reality of humankind's great togetherness with each other and with Earth. I will hold you in my heart from this day forward as a precious gift of awareness from the Great Atlantic Teacher.

Suddenly, I see Mt. Pelée as Earth's finger, pointing straight at me. Earth is seeing me as its reflection. We inter-are.

REFLECTIONS ON THE GREAT TOGETHERNESS

Days after I was in the vineyard and saw that nothing was separate, a kind nun told me the paperwork required for my ordination into the Order of Interbeing was lost. I smiled and reassured her that everything would be all right. How could it not be? A formal ceremony would change nothing—I already had a life-altering ordination on the hillside, witnessed by the old vines. My vision lasted for hours, faded slowly, and the ordinary presentation of the world returned completely by the time I went to bed. But the power of witnessing my seamless integration into the community of life didn't fade. I had swum in the ocean beyond my small-wave-self. I experienced my true belonging as a cell in the larger body of the human and nonhuman community. United, I felt whole and free and safe and enormously happy. What could an ordination ceremony offer that could top that?

"On the off-chance your paperwork shows up," she said, "why don't you write a note about your aspirations for practice?"

I followed her instructions, collected a pen and paper, and found a seat in the grass next to the hamlet's lotus pond. Bunches of flowers grouped together and reminded me of throw rugs strewn across a puddle. Sitting there, I thought back on battles I had waged in my personal and professional life. A question came to mind: if ultimately there is no other, then who the hell have I actually been fighting with all this time? Me? The larger body that is my own self? If so, isn't that a lot like my big toe waging war on my nose? Well, that's ludicrous.

Wouldn't it be better if I didn't constantly feel divided from those who give me grief, but approached difficult relationships instead from a place of unity? Maybe, I thought, the whole point of all this gnashing of teeth and beating of chests is that we're trying to wake up to our interbeing and we don't know how. Of course, I realized, I couldn't possibly get along with everyone—some people were completely welded to their idea of a separate identity. But I suspected many would very much like to stop feeling so alone; many would love to connect in a deep and nourishing way, discover themselves as a member of one beloved community hurling through the wilderness of space-time. Connected, I thought they too would find it simpler to give up warring and choose instead to take good care of each other as an ultimate act of "self"-love.

In that moment, my own inner turmoil seemed so unnecessary. It felt downright unhelpful, too, since my goal was to be a source of well-being. If I'm an integral part of the one beloved community and there's no separation between my heart and that of the larger body, I thought, then there could be no peace in the world until there was peace in me. Right? This meant I had to clean up what was in my heart. It is my job to take care of the space I'm responsible for in the cosmos. At once I deeply understood Thay's calligraphy: Peace in Oneself, Peace in the World. This is the fullest expression yet of interbeing. It is a reflection of our great togetherness with each other and Earth. So this, I decided, was my job: being peace.

In my best handwriting, I expressed my wholehearted aspiration to embody peace.

The next day, the nun approached me after morning meditation, "We have found your paperwork," she sang. "Thay has read it and everything seems to be in order."

"That's wonderful news, sister!" The two of us giggled with joy.

"Thay asked me only one question: 'Is she happy?' and I said, 'Yes, Thay, she is happy.' Because you know," she said in confidence, "Thay says the true measure of the quality of someone's practice is the level of their happiness."

The last day of the retreat, hundreds of monks, nuns, and laypeople sat on cushions set in close rows. The monastics were dressed not in everyday brown but in ceremonial saffron-colored robes. Spring air circulated through open windows bringing the freshness of the vineyard into the room. Fourteen of us were being ordained that day, and we sat nervously in pairs down the open center aisle. Together, the community chanted, recited the Fourteen Mindfulness Trainings, and we touched the Earth in full prostrations to demonstrate our commitment to study, practice and observe the trainings. Then Thay called each of us by name to come forward to receive the Dharma name he chose for us. When it was my turn, I stood and approached my Zen Master.

"Congratulations, Heather. Your new Dharma name is *True Lotus Peace*."

"Thank you, dear Thay," I whispered through the lump in my throat. I bowed deeply from the waist with my palms joined over my heart in the shape of a lotus bud. Hot tears fell onto my cheeks. Back at the cushion I felt love vibrating through the room—not in the goo form of my vision but in a way I would come to access often in the days and years ahead, during moments when I suddenly sensed the great togetherness of which I am a part. I knew those present in the room would continue in me. The vineyards of France would reverberate also in my precious awareness of interbeing.

During my time in Saint-Pierre, the Great Atlantic Teacher handed me a refresher course on the lesson of deep connection. No one and nothing was separate from the mountain the day Mt. Pelée blew. But I know where the leaders of Saint-Pierre were coming from in 1902. Like me, they had digested the views of their family, lessons from school and places of worship, along with cultural expectations from their work place and the ideas of their friends. They embraced the thrill of competition, took personal pride in their success above others, and were rewarded for their individualism. But by over-identifying with their separateness, they were like fish who thought they could swim on their own because they did not see the water (the bond of the community) that both united and sustained them as a member of the reef. Not once did they put their trust in the community with the level of confidence that a sea anemone holds for the coral on which it grows. When their way of life was threatened, Saint-Pierre's leaders doubled-down on their us-against-them mentality, and this cost them and 30,000 others their lives.

I've had us-against-them thoughts in my advocacy work too—moments when I didn't know or forgot about my great togetherness with the larger body. Trapped in the jaws of my own discriminating mind that divided self from other, I found it impossible in those moments to stop and consider the larger reality that each of us is woven into a cloth of one community. The mistakes these historical leaders made are my own, and knowing this stirs not blame but compassion in me. Valuing individualism and competition more than the collective and cooperation disrupts natural harmony and balance. This makes the wilderness through which we sail in our daily lives far more dangerous to navigate. And when we forget our great togetherness, we don't know where to turn when nature's balance becomes gravely out of whack and the environment starts to explode.

The Great Atlantic Teacher (with the help of Mt. Pelée) showed me that when we divide ourselves from others, we are capable of life-destroying actions. In contrast, the insight of our togetherness offers a radically different and healing perspective for civilization, an antidote to our separate-self-centered consciousness, a way of being and acting rooted in love. To put it plainly, the challenge before us is to see the ways in which we inter-are with people and the planet, put our trust in the wisdom of the human and nonhuman community, and listen to one another to figure out how to take care of the whole.

The discovery that we are in community is news good enough to bring profound relief. It means no one need face difficulties alone (especially something as intimidating as the climate challenge). What's more, there is a place for everyone in the beloved community. We need the wisdom of all the diverse cells in the collective body to know what to do. The reality of our connection is a new story for the whole of civilization, and operating from the wisdom of relatedness is a radical act. It is the stuff of peaceful revolution and lasting transformation.

Buddha was the first to suggest we could free ourselves from our discriminating minds (our habit of dividing self from others). It's time to wake up to our place in the one beloved community—human and nonhuman, living and nonliving. Everyone is welcome and included (and needed) as a thread in the collective fabric. With the insight of our togetherness, we can help one another navigate the wilderness of the cosmos. The consciousness of interbeing is the spiritual environment that can support societal transformation and Earth's healing. This can be the new story for civilization, one that (finally) works for all beings.

Here is a guided meditation I wrote and use to keep my awareness of interbeing alive.

Breathing in, I see myself as a cell in the body
of the beloved community.
Breathing out, I let go of my separate self-identity.
In, a cell in the body of the community.
Out, letting go of self.
(Ten breaths)

Breathing in, I see myself as a cell in the
larger body of nonhuman life.
Breathing out, I see myself harmonizing with nonhuman life.
In, a cell in the nonhuman body.
Out, harmonizing.
(Ten breaths)

Breathing in, I see myself bound in community to
those who are difficult or dangerous.

Breathing out, I extend compassion for their discriminating
    mind and notions of separateness.
In, one with difficult beings.
Out, extending compassion toward habits of discrimination.
(Ten breaths)

Breathing in, I listen to the wisdom of the
    human and nonhuman community.
Breathing out, I hear the wisdom of the human
    and nonhuman community.
In, listening to the community.
Out, hearing the community.
(Ten breaths)

Breathing in, I take refuge in the collective
    human and nonhuman body.
Breathing out, I trust the collective to sustain life.
In, take refuge in the collective.
Out, trust the collective.
(Ten breaths)

CODRINGTON
BAY

CODRINGTON

BARBUDA

ANTIGUA TO BARBUDA

ANTIGUA

ST. JOHNS

12 PERCEPTIONS

I can't see land. The island of Barbuda is flat compared to its mountainous neighbors like Antigua. I aim *Wild Hair* north, where nautical charts say the island sits, but there's only water—iridescent, peacock-feather-blue water. We've got to be close, I think. A calm blow pushes our ship across soft seas and every once in a while a finger of breeze tickles behind the jib to make it luff. But the next puff inflates the canvas faster than the throat of a randy frog and the sail complains with a mild snap as it fills. We've luffed and snapped for seven hours now in our voyage between the islands of Antigua to Barbuda—two halves of a single nation. I'm feeling mellow, but it is slightly unsettling to not be able to see the island we're sailing towards.

Maybe I shouldn't be looking for land, but for a small clump of trees at the tip of Cocoa Point. Once I can see trees, the guidebook says I should steer clear of the reefs near the surface and avoid being the latest entry in the registry of local wrecks. I reach for binoculars and scan the horizon. Nothing.

My eyes lift. A cluster of clouds hangs just ahead on an otherwise cloudless day; it's a cartoon bubble awaiting text. The clouds float because land somewhere underneath is warmer than the ocean's surface; heat makes evaporation, forming the clouds. So what I'm looking at is the island materializing in thin air. I heard once that ancient Polynesians were sky-readers; moving toward clouds is how they found Hawaii. But these puffs are unique, because they have pink underbellies. The red coral growing in the surrounding reefs turns the sand on Barbuda pink, so the rosy tone reflects in the billows overhead. What's underwater is up in the sky, too.

I pick up the bag of frozen corn at my side and press it against my swollen eye.

"Is that still cold?" Dave asks.

"Yeah. It feels good."

"Let me see." I lower the corn and turn his direction. "Your eyelids look like two pudgy caterpillars glued lengthwise. I wonder what the other guy looks like."

"The other guy is a mosquito, and no doubt he's fat and happy in the Caribbean sunshine today." Dave laughs. "The sneaky bastard went a little nuts with the anti-coagulant juice trying to suck blood from my lid, no?"

"That's a good theory."

Dave's right. It's just an idea. I remember again Thay's teaching on the nature of perceptions—how most of what I think I know is flat out *wrong* because there's so much going on in reality that I don't catch. As a human, my consciousness is distracted and jam-packed with thoughts and feelings, so I shouldn't trust my perceptions too much.

"It's a good guess," I concede.

"You look horrible. How does the rest of you feel?"

"The antibiotics are winning. I don't think the bladder infection will kill me."

"That's good. I'd miss you."

It's been a month of adventures. After leaving Martinique, we sailed to the Commonwealth of Dominica where, on our arrival, a not-so-helpful local in an inflatable boat skirted at the last second between *Wild Hair*'s bow and the mooring ball on which we planned to tie off. Sandwiched, I watched the local craft squeeze and twist in slow motion. *Wild Hair* was unharmed, but our bow snapped the arch holding the local's running lights. It cracked his outboard engine cover too. Then things sped up as Dave and the man launched into argument. The language that strung from the man's lips to my ears was fouler than one of Dave's sardine and ketchup sandwiches. Eventually, we offered the fellow $100 if he would sign a paper saying we had no further liability and suddenly we were best friends. Over the next couple of days, the guy spent the cash on the needed repairs (which he proudly showed us and explained in detail) *and* a blaze orange life jacket *and* blaze orange lettering that read "Search and Rescue Boat." During our entire stay, the fellow guarded *Wild Hair* like a fan guards fifty-yard-line seats at the Super Bowl. We left feeling genuinely good about helping launch the man's career.

The Commonwealth of Dominica was spectacular. At Trafalgar Falls, we swam all by ourselves in and around natural pools of different temperatures (from very hot to chilly water). The ponds littering the valley floor appeared to be decorated theme-park style (with boulders and exotic plants) and the water temperatures varied dramatically because the twin waterfalls that cascaded down the mountain to feed the ponds were from different origins: a cold mountain lake and a scorching volcanic stream.

One day we sat at a street-side café as Dominica's rowdy Mardi Gras parade sashayed by. Another time, I went scuba diving and swam through seeping underwater volcanic gas steaming upward like hot champagne bubbles. On the north end of the island, a guide rowed Dave and me through dense jungle, up the Indian River to a grapefruit plantation. When the sun dropped, we enjoyed a beach barbeque with other sailors in the bay, people who had ventured over from Europe.

We departed Dominica and sailed north to Les Saints and then to Guadeloupe. A storm blew past after we anchored *Wild Hair* in the Jacques Cousteau Marine Park. When it was time to go, we discovered our anchor dragged and got snarled in a mound of cables that were embedded in a concrete block the size of a car buried in the ocean floor.

Again, I suited up in scuba gear, sunk forty feet to the bottom, studied the tangle, and picked and sorted my way through the mass to free the anchor and us from the bond. That'll likely be the closest I come to feeling like an astronaut on a spacewalk.

Guadeloupe is shaped like a butterfly, and where the insect's body should be is the River Salee. It's navigable. Dave and I woke well before dawn to drive *Wild Hair* past the two bridges (obstructions that let boaters through only once a day and at a time before commuters wanted to get to work). Our guidebook suggested we should get beyond the bridges, toss an anchor, and go back to bed, saving the rest of the journey through the spectacular mangrove swamp for daylight. That was our plan. But, after we dropped anchor, I was restless and couldn't sleep. Finally, the sun came up and I looked around to notice our entire stateroom was full of no-see-ums; thousands of biting insects that had wiggled through the hatch screen to form a haze above my body.

"Dave, we're being eaten alive!" I shouted. My man hopped out of bed and within seconds *Wild Hair*'s engine was on and I was at the bow pulling up anchor. The problem was, the swamp bottom was all muck. In my frantic dance to fend off the cloud of blood-sucking insects, I was simultaneously pushing the foot pedal of the electric motor to draw up the anchor chain and spraying the muck with a hose. As I flailed, water shot everywhere. Once the anchor was out of the water, I signaled to Dave that *Wild Hair* was free, turned to go back to the cockpit, and saw Dave clomping and whacking frantically in his own bug cloud. Did I look that silly? Luckily, once *Wild Hair* got moving, we outpaced the horde.

Later that afternoon, though, I went into the master stateroom and discovered the bugs riding as stowaways. A can of bug spray put swift end to that. Later still, when it was time for bed, I discovered our bed sheets coated in a half-inch of deceased insect carcasses. In no time I stripped them off and shook the whole lot into the drink.

Nature can be a little overwhelming sometimes.

Antigua, the mega-yacht capital of the Caribbean, was a glamorous place. But Dave and I were especially fond of Great Bird Island—an undeveloped haven perfect for hiking and snorkeling that is part of the barrier chain buffering tall peaks and big cities from the full effects of the Atlantic. We had the anchorage out there all to ourselves. Back on Antigua a couple days later, I needed medical attention for my infection, and the doctor stunned me by telling me he and his family *owned* Great Bird Island (along with *six other* Caribbean islands). Holy smokes. What does one say to a thing like that?

Now we're nearing Barbuda.

"Do you need anything?" Dave asks.

"I need to see land before we get too deep into the reefs."

"It's weird, no? There are all these volcanic mountain islands dividing the Caribbean from the Atlantic, and then this flat little coral atoll all by itself. What's that about?"

I shrug. "Hold on—is that what I'm looking for?" I point at what appears to be dark sticks poking out of the water. Dave turns.

"I guess so. It looks like you just located Cocoa Point, baby."

"Good grief, this place is flat." After a minute, several buildings and a water tower rise like a sea serpent from the depths. Their presence is the only indication of an island fourteen by eight miles long. I adjust *Wild Hair's* course toward Palmetto Point and our final destination for the day off Eleven-Mile Beach.

This island has a different history from the rest of the Caribbean chain. In 1674, it was a British colony. Sir Christopher Codrington established the largest sugar plantation in the islands on Antigua, next door. To power his agricultural production, Codrington leased Barbuda from King Charles II. According to historians, he used the island to raise slaves and supplies. The Africans of Barbuda couldn't escape, but they were mostly left to their own devices. They were self-governing subsistent farmers and fishermen that were independent thinkers, and when Britain abolished slavery in 1834, the residents of Barbuda refused to leave.

Sugar was never grown here. So the heavily wooded landscape of native scrub, palmetto, and mangrove remains to this day. But what's really unusual is the island's donut shape. At thirty-seven square miles, the barely contained Codrington Lagoon runs the length of Barbuda's western coast. On the chart, the island looks as substantial as a hairband with a frill. The northern tip is a biologically prolific tidal marsh ecosystem. Baby lobster, fish, and endangered Hawksbill and Leatherback turtles thrive in the seagrass, mud flats, and coral reefs. The most uncommon creatures here are Barbuda's Magnificent Frigatebirds—giants who nest in the local mangroves September to April.

The hole in the donut, Codrington Lagoon, is the centerpiece of Barbuda's financial and cultural stability: it is both the hatchery supporting sustainable fishing and the avian nesting ground drawing tourists. The wetland is so important that, in 2005, Antigua-Barbuda entered into contract with the international Ramsar Convention to put almost 9,000 acres off limits to development. They joined the convention to safeguard not just wildlife habitat, but things like nutrient cycling, flood control, and other economic, cultural, scientific, and recreational assets. Meeting those lofty aspirations requires scientific monitoring and sophisticated resource management strategies.

Wild Hair coasts along the western edge of Barbuda, and I see nothing but shimmering pink sand. The sun, which shines from behind and over my shoulder, acts like a torch illuminating coral heads, some of which are only inches beneath the surface. I don't want *Wild Hair*'s keel to hurt the reef (a fragile living system that grows only an inch each year) or, for that matter, the reef to crack into *Wild Hair*'s keel. So Dave is standing watch on the bow. Every once in a while, he raises an arm in the direction I should turn. I zig-zag *Wild Hair* slowly and deliberately through an underwater maze. Then, right in the middle of the longest and most idyllic stretch of beach I have *ever* seen, with no one and nothing around, Dave drops the anchor. It buries effortlessly into the deep sand.

"C'mon out, Dinghy," I call as Dave and I tidy lines and stow charts and gear. Her tiny face peeks through the companionway opening. Sleepy eyes blink in the brightness. She jumps onto the cockpit seat, extends her front paws, sinks claws into the seat, and arches her back—tail extended and twitching with the stretch. Then her head pivots as she licks her lips and takes in the new surroundings. It's no wonder Dinghy's veterinarian is jealous of her patient; the cat gets all the pleasure of cruising and none of the responsibility.

Dinghy jumps from the seat onto the deck and moves toward the mast. She stands on her back feet and sinks her claws into the ropes secured to the mast. Then, she looks up and scales the pole like a Marine—paw over paw. I don't know how she manages—her arm muscles look too small to lift her wide rump. When she reaches the mast's intersection with the boom five feet up, she stops. In a deft maneuver, Dinghy half-leaps, half-twists her way onto the boom and into the folds of sail cloth. Tiny paws creep deep inside the honeycomb of secret fabric rooms. Here she will sleep, with trade winds gently funneling the length of the sail, until sunset when Dave and I will have to prod the boom with boat poles to nudge her from her perch so I can bring her inside for the night. She will emerge near the intersection of the boom and mast, vocalizing complaints over being disturbed.

The vacuum in *Wild Hair*'s fresh water system is acting squirrely, so we're on water rations until Dave figures out how to get the taps flowing. I smear shampoo through my hair, soap my body, and jump into the bright blue bathwater of the surrounding sea. The warm saltwater feels delicious, but the view through my mask is milky. Particles of flour-fine sand float so densely that I can't make out my fingers or toes. Instantly, I'm unnerved and my imagination begins fabricating dozens of hungry sharks. I'm quite sure they're chomping their multiple rows of teeth and plotting a meal of ankle, thigh, or hip. The more I tread water, the greater my panic. Don't thrash, Heather, I tell myself. That only makes you more desirable. So in a single motion, I drop beneath the surface, swish my head to-and-fro to get out the suds, and climb out of the sea on the ladder mid-ships.

"That was fast."

"This beautiful blue is made of milk. I can't see what's coming, so I'm outta there." Using the half-bucket of fresh water and the cup Dave set on the transom, I ladle the salt from my hair and skin.

"Hmm. Milky, huh? I guess I'll be super quick, too. At least it's pretty to look at."

～～～

Now it's morning and I climb into the cockpit with coffee and a warm scone I just pulled from the oven.

"*Garden of Eden, Garden of Eden, Garden of Eden*, this is *Wild Hair, Wild Hair*." Dave is at the navigation station below, arranging for a boat tour to the Magnificent Frigatebird Colony. We hope to explore nature and then clear customs in the Village of Codrington out of Antigua-Barbuda in advance of tomorrow's sail to St. Barts. (We're still making up for the time we lingered in Martinique.)

Garden of Eden—that's a quaint name, I think. I stretch my arms overhead, and observe how the morning sun bastes the world in pinks and yellows. Light dew makes *Wild Hair*'s deck gleam. The hull gurgles on a vast sky-blue platform. A sip of hot coffee coats my throat.

I've read about this island and its people, how the bounty of nature has sustained local residents for years. Here, fresh water, seafood, livestock, and soil for farming are unlimited. There is wood for buildings and sand for concrete and mortar. Left to their own devices, residents organized schools, churches, and a society of democratic self-governance. I never imagined Eden as an oceanfront community with beaches as far as the eye could see, but why not?

I feel lucky to be here. Barbuda dangles at the far northeast corner of the Caribbean Sea. We almost couldn't come because of the weather. The forecast had to contain several days of mild conditions before we'd commit, and to our great good fortune, that's the prediction we got. Today there are a few sailboats anchored way far off, but here at Eleven-Mile Beach it feels as if the universe is ours and ours alone.

"We're set," Dave says, opening a bag of bread and speaking from below. "Patrick will pick us up at Louis Mouth at eight. He'll be our guide to the nesting area and our water taxi to town."

"Perfect."

I step from the dinghy, past the surf, and blush-colored sand swallows my foot to the calf. With the next wave, the inflatable knocks me to the ground, onto my hands and knees. I gasp in surprise, wiggle my digits. Barbuda's bizarrely squishy, cozy-warm, unsoiled sand is a tactile thrill. The ratcheting of the outboard as Dave tilts the propeller from the water refocuses my attention and I struggle to my feet to pull the inflatable up onto the beach to safety. When Dave jumps out, his feet sink and he tumbles.

"Careful, it's softer than it looks," I grin.

The two of us yank the dinghy up the dune, struggling through the *terra mollis*. We wrap chain around a scrawny tree growing atop the crest dividing the Caribbean from Codrington Lagoon. The land isn't much, only a thread of sandbar twenty to thirty paces wide. Patrick is nowhere in sight, so with the dinghy locked and safe, I wander. The interior water is dark, a soupy green. A little ways away someone organized a few logs into a circle of benches. There is an abandoned folding chair, a lost shoe, and remnants of a bonfire. It seems like a decade of beer cans and fishing debris litter the scrub. No doubt humans dropped an equal amount of litter on the Caribbean side of the sandbar, and the sea swept it onto the reefs and inside fish bellies.

I am confused by the mess. I don't know what to think. The people of Barbuda put this lagoon under conservancy with the Ramsar Convention. Additionally, residents once rallied and successfully stopped construction on a desalination plant. Another time, when the mobile office for the construction of a new hotel arrived, locals pushed it off the hill and into the ocean. These actions make me think no one is going to run roughshod over these people or this place. They are environmental advocates, my kindred spirits.

But even so, they've boxed themselves into something of an ecological nightmare. Tourism didn't grab hold here but *sand mining* did. These people are literally selling the land from under their feet, carving bites bit by bit from the body of the island and serving it to hotels and coastal communities worldwide. It started in 1976, when the central government of Antigua leased River Beach on Barbuda to an outside mining company.[35] Barbuda objected. After more than a decade of court battles, executives from the mining company went to jail and the mining rights reverted to Barbudans. But in 1993, rather than stopping the practice altogether, residents *took up* the practice. Now, mining continues and profits from the sale of this luscious sand generates a third of the Barbudan Council's revenue. It's how locals pay for roads, electricity, telephones, and healthcare. Scientists and engineers tell them sand-mining is an unsustainable enterprise. (Islanders have blown well past removing what anyone would consider to be a safe amount. Marine scientist Eli Fuller calls it a "destructive and irrational practice."[36]) But islanders claim they need the revenue and haven't found a way to stop. Today, even though sand mining is eroding any chance Barbudans have to adapt to sea level rise, the island's natural resilience continues to be hauled off, bucket by bucket, barge by barge.

I don't know what to make of these inconsistencies, and I'm trying not to leap to any conclusions. Thay says my perceptions—like strong emotions—are something I ought to question. They almost always have little to do with reality. Ancient wisdom says something like 99 percent of my perceptions are flawed, because either I can't see the whole picture, my sensory organs give me the gist of what's happening but not an exact photocopy, or I'm filtering what I'm seeing through judgments, emotions, and preconceived ideas. So because the data entering my consciousness is incomplete or skewed, I try to give people the benefit of the doubt and question my perceptions. I don't do it just to be nice; seeing reality without the distortion of personal filters is practical. It makes me more clear and effective in the world.

Thay says I should I think of myself as a lake with a glass surface, reflecting all that is without distortion. When there are no ripples, insight comes. (I know from experience I must be *really* calm to keep my inner lake flat.) So, when sitting, I focus on my breath, relax my mind and body, and look deeply at what *is*. When perceptions and stories about reality bound into my consciousness and start thrashing about in what was the calm pool of my awareness, I thank them for sharing, let them slip away, and continue to calm my internal lake and observe what's happening in the present moment.

Here I am in Barbuda, and I have all sorts of preconceived notions about local residents and their addiction to sand mining. I'm not proud of these thoughts that are thrashing like toddlers in a kiddie pool. The actions of locals strike me as cultural suicide, and they are triggering in my head a cascade of suspicions and judgments. (I half expect our tour guide will show up sporting two heads and a tail.) I am mindfully aware of a full blown prejudice trying to take shape, and that's why it's so important for me to let my thoughts and perceptions go, and pay attention to the living, breathing, in-the-flesh people I meet here.

I will do this even as my mind keeps landing on one of the Buddha's most gruesome teachings. Once, he told a tale of a young family—father, mother, and baby—who chose to travel across a large desert. The journey didn't go well and suddenly they found themselves in the middle of a wasteland without food. Surely all three would die unless the parents did the unthinkable: kill and eat their precious baby to sustain their lives. The mother and father were devastated; they wished *whole-heartedly* that they had considered the consequences of their choices earlier and made other decisions along the way. But, arriving to a place where they had no choice other than to die, they acted swiftly, choked down some of their child's flesh, and walked to safety.

This is a teaching about avoiding regret. The Buddha told this story to prod his students into deeply considering the consequences of their actions *in advance*, while time remains to make different, sustainable choices. He wanted to expose the link between our unconscious decisions and the resulting heartbreak and suffering they often bring. To avoid the regret of "eating our children," he said, we can be mindful of the true cost of our actions. This grisly tale has inspired generations of followers to reflect on the effects of their actions and *change course* in order to safeguard the future.

How many nights do I lie awake, grieving the fact that my generation is eating up the next generation's opportunity to thrive? I know I'm not alone, although it's difficult for ecologically concerned people like me to speak intimately about our deep suffering around society's unsustainable actions, about our regret and sadness. Everywhere, we see evidence of society eating our children and we are mortified to our bones. I think Aldo Leopold (the father of America's Environmental Movement) described it best when he wrote:

> One of the penalties of an ecological education is that we live alone in a world of wounds. Much of the damage inflicted on land is quite invisible to laymen. An ecologist must either harden his shell and make believe that the consequences of science are none of his business, or he must be the doctor who sees the marks of death in a community that believes itself well and does not want to be told otherwise.[37]

I scoop two handfuls of sand and toss the grains downwind. The powder gets caught by the breeze and sparkles as it soars, arresting my breath. This stuff is gorgeous. Will selling it today to luxury resorts and beach communities bring an end to their children's future on this island? At least one Caribbean marine biologist I read about thinks two-thirds of the landscape is already poised to become wetlands; when it does, just twenty-one square miles will remain to sustain the population.[38] Moreover, it's likely more intense storms will resculpt the territory. This already happened once, when Hurricane Luis (a category four hurricane that plowed full force into Barbuda in September 1995) chewed the shore inland by sixty-six feet. That day, the narrow sandbar I'm standing on now breached in several places. Had we arrived on Hurricane Luis's heels, *Wild Hair* could have sailed inside the lagoon.

It seems to me the marks of death are all over Barbuda. Is this a perception I've got wrong? I hope so.

A Trinidad-built twenty-five-foot open skiff whizzes toward us from the village on the far shore.

"Good morning, good morning!" Patrick says, arriving. He's fifty-something, handsome, and strong with only one head and no tail (as far as I can tell).

"Are you ready to see Barbuda's Magnificent Frigatebirds? They are ready to see you! Please, step aboard." I smile in return and take his extended hand for balance as I board.

Dave and I settle on seats that could accommodate a crowd. Patrick reverses the boat's massive outboard, then accelerates forward. Mangrove clumps lying to the north sit upon the surface. Thirty feet overhead, black silhouettes—prehistoric crisscross shapes looking like flying capital *M*'s—swoop. Their split tails trail like tuning forks.

"Are those the Frigatebirds?" I ask.

"Those are the female *Fregata magnificens* looking for husbands. The males roost on the nest until a female approaches to be their companion. It's ladies' choice."

"How nice!"

Patrick stops the engine and lifts a long stick from its resting place on the gunnel to pole the boat through a narrow channel. He's a Barbudan gondolier. More sticks driven into the muddy bottom mark a channel.

Noise and a sour, gamey smell overpower my senses first when we turn the bend. Then the world is in chaos. Mangrove bushes squirm with life. Bills clap rapid fire: *bap, bap, bap, bap, bap.* Birds squawk and scree. Fuzzy white faces, hopelessly comic in their confusion, stare our way. Giant red pouches swell from the necks of the otherwise black males, inflating the suitors to twice their size, advertising virility and desire. Flying females circle close enough to touch. Ladies are landing on mangrove branches. They hop and leap and seem tormented by indecision, wondering, "Who do I choose, who do I choose?"

"When a lady Frigate comes near, the male begins his courtship," Patrick explains. "Watch how they wag their heads and shake their wings. The clatter of beaks is very attractive to the ladies."

Bap, bap, bap, bap, bap, bap, bap. Another neck billows into a bright birthday balloon. *Chip, chip, chip.*

Screeeeee.

I feel myself falling deep into the source of mystery and miracles, as the organized chaos of a foreign society negotiates its continuation without a care to my presence. These bizarre creatures are experts at being themselves.

After many long moments, I pick up patterns in the Frigatebirds' colony life: the way they collapse their six-foot-plus wingspan to land precisely where they aim, how the mother regurgitates food to feed the chick, the manner in which the father sleeps in the branches by extending his wings like an open umbrella and dropping his head toward darkness beneath the leaves.

My eyes keep returning to the babies, who look like cartoon plush toys (with black unblinking plastic eyes) set upon a shelf. A chick turns its head and the stuffed animal illusion evaporates. One girl opens and closes her beak in silence; perhaps she's testing new jaw muscles. Or maybe she's signaling hunger.

"Both parents incubate the egg and then care for the chick," Patrick says. "The father may leave after three months, but the mother will stay for up to a year feeding the baby and teaching it how to fly."

"Typical male," Dave mutters, smiling.

"This nesting colony is significant?" I ask.

"Yes. This is one of the largest in the world, with about 5,000 birds. When a young bird becomes ten years old, it will return every other year to its hatch site to breed."

"How long do they live?"

"No one knows for sure, perhaps forty or fifty years." The realization that some of these birds are my age stuns me. They have amassed a long lifetime of experience. I try to imagine what other skills they've honed that aren't obvious in this moment.

"This is the Man O' War bird. They eat fish, but they are poor fishermen, so they fly together at forty-five miles per hour, attack other birds and make them drop their catch. Frigates are very aggressive and augment their diet of flying fish with food caught by other birds. The official term for this feeding is 'kleptoparasites.' It's the reason we call fast and maneuverable warships frigates."

"That's fascinating," Dave says. "I had no idea."

"They are excellent air acrobats. A Frigatebird can catch the meal released by another seabird before it drops into the ocean. They can never land in the water. They have no oil, so water soaks their feathers and it becomes too heavy for them to lift their bodies out of the sea. If they make a mistake and end up floating, they die."

"Oh, no. I thought they could dive for fish like loons."

"That's a common misperception. Once, my friend saw a Frigate land on water, and right away her two friends came. Each one took a wing on their beak and lifted. With that, she flew again. They saved her life." I let the intelligence behind that act sink in.

"They are all wing," Dave says. I look and see that the bird's body can't be more than a couple pounds.

"Yes. Their feet are small and they cannot walk on land. They only touch down in these bushes. But their light bodies let them soar in warm, high breezes for up to twelve days."

The Magnificent Frigatebird population is rising—conservation efforts on islands like Barbuda are succeeding for now. But as sea levels rise, low-lying mangroves will submerge and the loss of nesting colonies will make the birds vulnerable to extinction. My hands squeeze my knees as I worry over my new friends' future.

The Magnificent Frigatebird who greeted me two-thousand feet in the air before we made landfall in St. Thomas comes to mind. He was the first life form I met after weeks at sea and he was hundreds of miles from shore. His presence was a sign, a harbinger of land, a deep comfort, and now, I have a much better sense of his essence. I wonder if the fellow is here, in front of me.

Patrick poles our boat a little deeper into the reserve. It's difficult to imagine we evolved from the same brine, that the birds and I are cousins. I study one lady's beak—a long candle with a waxy end that's melting at the tip—and wonder about the evolutionary advantage of the hook. I fantasize about the beak's texture, what it would be like to slide my fingers the length of their bill and push the tip of my thumb onto the point.

"It's the mothers who teach the babies what to do with their big wing span. Until then, the chicks sit for months and look perplexed."

"Can we go further into the aviary?" I ask. "It looks like it goes for miles."

"Yes, the nesting ground is very large, but protected. We can only go this far so we do not disturb the birds." My lips involuntarily sag. "It is very important that Barbudans respect this boundary. We used to eat Frigatebirds, but now we have stopped. The animals are safe from us."

"Have you eaten them?"

"Yes. They are the consistency of chicken, but taste like mild fish. Their meat is very delicious." He smiles and shrugs. "We have a lot of wildlife on Barbuda," Patrick says. "The West Indian Whistling Duck, wild deer, boar, donkeys, horses, pelicans, osprey, hummingbirds, the Barbuda Warbler, which exists only here, and nesting turtles, too. But most people want to see the Magnificent Frigatebird because nesting grounds are rare, and this is one of the largest."

I realize I've snapped a hundred photos of large birds and their offspring flopping in trees. I can't help it. The green, black, red and white scene is a living lava lamp and something I want to hold onto. Slowly Patrick poles the boat back up the channel. I could spend the day looking, but the tour is over.

Our craft arrives parallel to the village pier, and Patrick motions to two boys perched on a rail. The fellows grab lines and secure the boat. Words are exchanged between the generations in the local Patois dialect.

"This is my grandson, Wayne, and his friend, Devon," Patrick says. "They will guide you to the customs office." I imagine the boys have things to do on a Saturday other than show us around.

"I'm sure we can find it on our own. We have a map."

"No, they will be your guides. I am going to take another group to the mangroves, but I will be here when you get back to taxi you to your boat again. We will settle up then." I catch on that Patrick is not asking but telling us the plan.

"What grade are you fellows in?"

"Sixth." Wayne looks at his feet and Devon toward the sky.

"Sixth grade is a good age. Let's get going!" I smile.

We walk the paved road toward the island's interior. The one-story stucco house to our left is a freshly painted beige with light blue hurricane shutters. The yard is simply but purposefully landscaped. Electrified wires hang overhead. On our right, a brick half-wall shields the neighbor's garden. A bicycle leans against the wall, lock-free. A line of tiny homes stretch single file. Trees are few, but where they do stand, three or four chairs sit vacant in thin shade. Up the road we pass a cluster of one-room buildings painted in the soft colors of a sunset.

"That is my school, Holy Trinity School," Wayne volunteers, "and this is the government office, the Barbuda Council." His left hand makes an involuntary gesture. Wayne is a shy boy—I hadn't expected conversation. "Across the street is the Post Office." His right hand twitches.

"You know so much. Will you give us an official tour? I would love to know the town as well as you."

"OK." Wayne's face bends into a sideways smile. His eyes peek my way. Devon walks several steps behind.

"And Devon, you can explain things too, if you like." The smaller boy darts his eyes to the ground, bites his lips, and swoons slightly.

"If you want to rent bikes, they have some up here."

"Next time for sure."

We continue walking and enjoying a mix of silence and conversation. Wayne identifies the island's food market, phone store, police station, and restaurants. We turn a corner and suddenly we are amid dozens of people milling in the street; everyone is carrying trash bags. It's a neighborhood cleanup involving what feels like the whole village. A very large man steps toward us carrying a half-full trash bag and wearing blaze orange gloves.

"Good morning. Can I help you today?"

"Ah, no, we are all set. These young men are taking us to the customs office."

"OK. My name is Matthew and I am the customs agent. I will meet you there in five minutes."

"Thank you," I stammer, surprised. "That would be very nice. We will see you there."

"Good morning," a woman says, picking up a soda can. "Good morning," three more cheer from across the road.

"Good morning, good morning," Dave and I say in unison. I can't remember being in a community of strangers who expressed such an easy kindness. With the way these people are picking up trash, I'll bet the trash-laden sandbar of Eleven Mile Beach is next on the list.

There are as many bicycles as cars on the island, maybe more. People get around on foot, too. At the next intersection, a man parts company from his friends to come talk to us.

"Good morning," we say. He falls into step with us.

"Very soon now," he starts, "visitors will stop going to Antigua and will come here instead. This is the place to be—it is so much better!" He speaks as if we're picking up from a past conversation.

"I see." I say. The man is tall, slight, and neatly dressed.

"All we need is a really good port and people will come here."

I can't think of anything to say. "We're on our way to customs and these fine young men are showing us the way."

"You're in good hands then. They will take good care of you." The man continues walking in silence next to me.

"When will you be leaving the island?"

"Oh, we have to leave tomorrow, early in the morning."

"That's too bad. Tomorrow there will be horse races after church at four o'clock. I have a horse in the race I've been training: a chestnut thoroughbred named 'More Blood.'"

I stop in my tracks. Horse racing in Barbuda? What an unexpected past time.

"Does More Blood win?"

"Oh yes," he says happily. "He's a little big right now, but I've been working with him to put him in shape."

"Do you do the racing?" A full belly laugh shakes his shoulders and limbs.

"No, I ride him, but I have a jockey for the race."

"A small man?"

"Yes, a small man to be the jockey." We continue walking. At the next corner the man continues straight as the boys swing right.

"Good luck tomorrow," Dave says.

"Thank you. And if you come, look for More Blood."

"We will try," I say, "but I don't think we will be there. We've got to get back to the states, but it was really nice of you to invite us. Thank you. Good luck!" The man walks backward and smiles before turning and continuing with his day.

"I'll bet if we stayed here more than a day or two, our dance card would be full," Dave says.

"You're not kidding. These folks are full of surprises."

Wayne and Devon stop at a small cinderblock and plaster home next to the airport. This is customs, I suppose. Matthew is inside readying the paperwork. Dave takes five Eastern Caribbean dollars out of his wallet (the equivalent of about two bucks) and gives it to Wayne. The boy looks unsure if he should take it.

"This is for you both; you earned it. You spent time walking us here and giving us a nice tour. You fellows buy some ice cream or something. OK?" The boys look at one another and then nod. Wayne's shy hand reaches for the bill.

"Thank you for your help," I say. We step inside.

The room is very dark. Matthew sits at a desk behind towers of papers; the stacks tilt on the desk and floor, leaving barely enough room for business. Matthew hands Dave several pages in exchange for our passports.

"So it sounds like the people of Barbuda don't want big hotels on the island."

"That is not true. We want the local people to benefit from development."

"I don't understand. I heard about the people pushing the construction container into the ocean."

"Yes. That I agree with. Those developers didn't have permission or a permit. They just brought their mobile office trailer and started fencing."

"What?"

"The land belongs to the people of Barbuda. We own the land in common. We decide together what happens here."

Before coming, I read online about the Barbuda Land Act of 2007. The instant the people of Barbuda were freed from slavery, residents claimed their right to the land and self-determination. Many years later, the Land Act came about to legally protect the rights of all Barbudans and their descendants (family members living at home and abroad) to build a house, farm, or use property to start a business at zero cost (as long as they follow the community's established regulations). In this way, the island is used freely and in common by the people. Foreigners can lease a parcel of land, and there are people living and working on the island from Europe, the Middle East, America, and other Caribbean nations. But the Land Act—an extraordinary piece of legislation—gives Barbudans legal means to conserve what other islands have already lost: pristine habitat, diverse wildlife, and a closely-knit community.

"Barbudans want honest partners," Matthew explains, "who are interested in sustainable, ecofriendly businesses. Developers have to bring funding to the table and not use our land as collateral. Small-scale ventures are right for here. Big development schemes? They can go elsewhere."

"That makes perfect sense." I want to ask why islanders continue to mine sand if they are so ecologically concerned. I can't think of a way to ask without sounding rude. I bite my lower lip. Dave finishes the paper-work and passes over the papers. "OK," Matthew says. "It's Saturday and the Port Authority is closed, so you don't have to stop there. I will call the Immigration Officer so she knows you're coming. Walk back to the pier and you'll pass the office on the way. Don't leave without clearing."

The Immigration Office is a tiny, mint green building with a gray flagstone façade. The place is locked tight. A hand-written note on the door reads, "Closed. For assistance, call 225-1265." Knowing Matthew phoned ahead, we wait. Thirty minutes pass before we decide to try calling for ourselves.

"What? You are there? Wait there—I'm coming now." Within

minutes a pretty young woman arrives by car. Her form-fitting t-shirt reads, "Just Hug it Out." She's flustered. Keys and bracelets jingle as she unlocks the door, flips on lights, steps behind the counter, and drops a load of books and files.

"How many more boats are out there?"

"We are maybe one of five, I'd guess." Her eyes roll and she takes a breath. "I didn't talk to Matthew or know you were here because I was on an international call. I came as soon as I knew."

"That's OK. Is everything alright?"

"Yes, fine." She relaxes as we fill out the forms, growing more companionable by the minute. It feels like I'm meeting an old friend, and I have an urge not to go but to pour a cup of tea and stay for a long while. Horse racing tomorrow is sounding better and better.

Eventually, we walk to the lagoon and meet up again with Wayne, Devon, and a third school chum at the public pier. Devon is worrying over the bait at the end of a hand line.

"What are you men fishing for?"

"Crabs," Devon says. It's his first word to me. Perhaps he's decided we're old pals.

"Are you catching dinner?" I ask. Devon nods. "Are you going to cook them, too?"

"No," he says, looking at me straight and serious. "My mom is a really good cook." Reverence and respect color his voice.

"Then you are a very lucky young man."

"Yes, I am." He is painfully sweet. I can't help but think of the day when he will have to leave this place because Barbuda's sand mining ate up his future. I suspect he'll be among the first needing to find refuge on higher ground.

Some believe there are alternative sources of revenue for the people of Barbuda: the government could explore sport fishing of all kinds or entice small cruise ships to anchor offshore. If islanders organize themselves to provide services and entertainment to passengers from a few ships each week, they might realize even more employment and tax benefits than mining now provides.[39] I can't understand why they continue to mine and destroy their children's future.

A crab pinches Devon's bait, and he lifts the creature onto the dock. He looks at me. I take in the boy's eyes, bright and brown. His smile is fresh.

"Look what you did! Your mother is going to be so pleased."

Patrick's fishing boat speeds our way, slows, and then stops alongside the public pier. Dave and I pay our bill on the taxi ride across the lagoon back to Louis Mouth. From behind the sandbar, *Wild Hair*'s mast points straight toward the clouds. Dave and I drag the dinghy down the Caribbean shore to the water, and motor home. Climbing the boarding ladder, I look up and see four horses from the ecolodge up the way carrying riders through the surf of Eleven-Mile Beach. The sun's late afternoon rays intensify the glow of rosy sand.

Part Two: Our House, Fitchburg, Wisconsin // July 2012

I enter the front door, carrying a water bottle, keys, and today's mail. The elastic bike shorts grip my thighs and the padding in my seat makes me waddle. It's hurricane season, so we're home again, enjoying the beauty of a Madison summer and catching up with family and friends. *Wild Hair* is double-tied to a marina pier in Georgia. I drop my load onto the counter, take a final swig of water, and thumb through bills and junk mail. Clunks and scrapes echo from the garage as Dave stores the bikes to his liking. Dinghy jumps onto the counter and rubs my arm with her shoulder. Her loose fur sticks to my sweat. Before I shower, I give the front page of the Muir View—a newsletter from the local Sierra Club chapter—a quick scan. The table of contents reads: "The Sand Rush (Fracking in Wisconsin)." My breath seizes.

On pages eight and nine I read that, although my state doesn't have natural gas or oil in shale deposits beneath the surface, our hillsides are made of a very particular kind of sand the fracking industry depends upon: the high-quality silica oil and natural gas companies inject into underground crevices to wedge open the cracks and get fossil fuels flowing. My stomach turns as I sink into the bar stool. Dinghy lies down and rolls onto the mail, wanting a pet.

A single fracking well uses nearly three-million pounds of sand during its operation, and 75 percent of the sand comes from the Midwest. The grains are strong and shaped just right along the Mississippi River. The demand for sand came upon the state so fast that no one was ready to regulate the industry. Now, as Wisconsin's hilltops disappear, so do rural ways of life, wildlife habitats, and air and water quality. With the largest frac-sand deposits in the country, Wisconsin is experiencing a great "sand rush." There are now 141 frac-sand mines leveling hilltops in my state. About nine thousand truck loads hauling Earth from beneath our feet every day. All of that is horrible, but it seems like the article is burying the lead: frac-sand mining is part of the fossil fuel industry, which is releasing carbon into the atmosphere and causing the climate to warm. We, the people of Wisconsin (like the residents of Barbuda but on an even larger scale) are devouring our children's future for short-term profits.

My arms and hands start to sweat anew. I was under the impression that Wisconsin wasn't involved in fracking. I cannot imagine why Wisconsin lawmakers are supporting this mining, especially since there are more sustainable development options for rural communities. Rather than turning the natural beauty of the state's bluffs and hillsides into deep sand pits, tearing apart the agricultural fabric of local communities, destroying ground and surface water resources, and enabling our national addiction to fossil fuels, Wisconsin could invest in green energy jobs, increase rural tourism, and promote sustainable organic farming.

I close my eyes and picture the rich farmland and oak forests I just biked through. I haven't advocated for anything since I took time off to sail. In four-and-a-half years, I haven't once called a meeting, framed a case, negotiated an agreement, or turned to the media to mobilize the public. During that time, I experienced healing: I no longer need medicine for a bad stomach or an irregular heartbeat. Now, I wonder, is it possible to engage with the thorny issues of our time in a different way?

Sitting at my kitchen counter, I decide to take a stand against the Wisconsin mining laws harming my children's future.

REFLECTIONS ON PERCEPTIONS

A handful of us sat around a large table in the community room of an ecofriendly cohousing apartment complex in Madison. The group included representatives from the Wisconsin Faith Voices for Justice, the Wisconsin Interfaith Power and Light, and my own meditation community: SnowFlower Sangha.

"Is that it then?" I ask. Five faces look at one another.

I had to do something about frac-sand mining in my home state, but what? As soon as I started investigating the issue, another local mining disaster materialized. The proposed Gogebic Taconite Mine in Wisconsin's pristine Penokee Hills threatened to obliterate Chippewa treaty rights and inundate sacred wild rice beds (and the expansive wetland water source supporting Lake Superior) with incalculable amounts of sulfuric acid. It was another idea harmful to people today and in future generations. Even so, the majority in state government flocked behind the mine because of the demand for rural jobs at all costs. Many environmental organizations were knee-deep in the mining science behind both frac-sand and taconite. Others were pacing the halls of the state capital trying to slow a political system hell-bent on extraction. But no one—as far as I could tell—had articulated the state's ultimate ethical responsibility to care for citizens, our nonhuman neighbors, and those who will follow us.

The environmental justice organization I led for years was up to the gills in other projects. So, a group of us in my local Buddhist practice center started talking, and soon the Earth Holder Project of SnowFlower Sangha formed. The *Earth Holder* name had meaning. It came from an ancient Mahayana Buddhist text, the Lotus Sutra, which mentioned a historical figure (the Bodhisattva Dharaanimdhara) who held, protected, and preserved Earth. (Thay wrote about this person in his book *The World We Have* and encouraged all his students to become "Earth Holders.")[40] As an expression of mindfulness practice, the Earth Holder Project met regularly to explore the link between our inner and outer sustainability and to offer ourselves and others opportunities to take skillful, Earth-healing actions. Eventually, in what was an election year, we joined with partners from other faith traditions and wrote an online petition urging the Wisconsin candidates for governor to end both fracsand and taconite mining.[41] Sensing the potential of a united voice to make a difference, nearly 1,000 people had signed on.

"Is there anything else we can do?" I asked the group again.

"We've spread the word to all our members," said a young woman from Power and Light. She was a gifted political thinker. "It's pretty clear that the election isn't going to swing one way or another on the mining issue."

"That's why we've stopped getting press on this," her cohort said.

"Well, we never really thought the election *would* turn on this," said the IT master from my sangha, "But every time someone signed the petition, the message went straight to both candidates' inboxes."

"And there was a bit of a buzz on the topic in the *Weekend of Coordinated Service*. Several congregations heard about the issue from their spiritual leaders," said the rabbi.

"Well, OK then. I think this was a good way to begin introducing the ethics common to all our traditions into a discussion dominated by politics, science, and economics. Do you agree?" Heads nodded, and we busted into the platter of bagels in the center of the table.

Ours was a small project when it comes to the amount of money and influence swirling around Wisconsin's mining deals. Still, months later, citing vague concerns over issuing permits, Gogebic Taconite LLC (the company at the heart of the conflict) closed their offices in Wisconsin and sent an application withdrawal letter to the Department of Natural

Resources. Today, an entire mountain once patrolled by armed guards and slated to be mowed flat has reopened for public and tribal use. What's more, although frac-sand mining continues in rural Wisconsin, the industry and demand for sand shriveled with the price of oil. The good news is we're making headway in my state when it comes to harmful mining (and I should say that not *all* mining is harmful).

It turns out, however, that we haven't accomplished nearly as much as the 1,200 residents of Barbuda. Like a lot of people, I followed the news online coming out of Paris from the 2015 United Nations Climate Change Conference, known as COP 21. One *Washington Post* headline grabbed my eye: "How tiny islands drove huge ambition at the Paris climate talks."[42] I clicked through and couldn't believe what I read.

Delegates from Antigua-Barbuda were at COP 21, and they were working the conference with the aggressiveness and precision of Frigatebirds in flight. Nations arrived with promises of CO2 emission cuts their countries were willing to take; the starting cuts, in total, translated into a 2.7° Celsius rise above preindustrial temperatures. The Antigua-Barbuda delegation sensed the 196 nations were feeling pretty good about themselves. With absolutely no political or economic power to speak of, the tiny island state pled for people to *get real* and protect Barbuda and other low-lying countries from rising sea levels. In written testimony, speeches, behind closed doors, tweets, and one-on-one appeals, they declared in no uncertain terms that even the 2°C increase that the most powerful nations were hoping for would be too much.

Antigua-Barbuda had misperception-busting science up their sleeve. They were members of the Alliance of Small Island States (a coalition that had pushed for a 1.5°C limit since 2008) and the group had done significant research figuring out the difference in impact between a 1.5° and 2°C increase: they studied things like the amount of sea level rise, the degree of stability of the Greenland ice sheet, the survival of coral reefs, and the vulnerability of global food security. The difference they demonstrated was substantial.

Beyond the science, Antigua-Barbuda led an ethical campaign supported by the Vatican, Maldives and a handful of other low-lying countries in Africa, Asia and Latin America. Their message: if Earth's temperature exceeds a 1.5°C increase, many defenseless nations (who had actually *not* contributed in any substantial way to planetary warming) would be condemned to death by drowning. The combination of science *and* ethics couldn't be ignored. Europeans, Americans, Canadians, and more African nations saw the reality of the situation and joined the campaign. In the eleventh hour, new language was written into the agreement and Antigua-Barbuda and their allies succeeded in wooing the world's super powers into a far more ambitious commitment. Although the COP 21 agreement officially contains the warming to levels "well below 2°C," it also clearly voices a collective aspiration to keep the temperature increase below 1.5°C.

Little Antigua-Barbuda—you surprise me. You sell the ground out from under your children's feet, and then you take flight and protect the future of all children everywhere. From my limited view, I could never have guessed that you would be so skillful in helping the world break through global climate misperceptions. Has the burden of sand mining ultimately inspired you to push for sustainability? Perhaps the teaching offered by Sōtō Zen monk Shunryu Suzuki is true for you: "Hell is not punishment; it is training."

If I were back on the island, I would thank every resident for protecting the Garden of Eden for us all. You are true children of the Great Atlantic Teacher. Your actions helped me understand more completely the nature of human inconsistency during this time of climate transition. You showed me the seed of transformation is inside every one of us; we each have the capacity to see reality clearly, and help others see it too. Now I know there is danger in judging others solely on past performance. Because of you, I will redouble my effort to guard against wrong perceptions by regularly asking, *Am I sure?*

To regain sovereignty over the tyranny of wrong perceptions and to help myself break through false ideas, I wrote and practice the following guided meditation.

> Breathing in, I am aware of my breath.
> Breathing out, I relax into gentle breathing.
> In, aware.
> Out, relax.
> (Ten breaths)

> Breathing in, I am still water reflecting.
> Breathing out, I smile to still water reflecting.
> In, still water.
> Out, smiling.
> (Ten breaths)

> Breathing in, I notice perceptions entering my
>   consciousness through my ears.
> Breathing out, I question my perceptions.
> In, listening.
> Out, am I sure?
> (Ten breaths)

> Breathing in, I notice perceptions entering my
>   consciousness through my nose.
> Breathing out, I question my perceptions.
> In, smelling.
> Out, am I sure?
> (Ten breaths)

> Breathing in, I notice perceptions entering my
>   consciousness through my tongue.

Breathing out, I question my perceptions.
In, tasting.
Out, am I sure?
(Ten breaths)

Breathing in, I notice perceptions entering my
    consciousness through my skin.
Breathing out, I question my perceptions.
In, touching.
Out, am I sure?
(Ten breaths)

Breathing in, I notice perceptions entering my
    consciousness through my eyes.
Breathing out, I question my perceptions.
In, looking.
Out, am I sure?
(Ten breaths)

Breathing in, I notice perceptions entering my
    consciousness through my ideas.
Breathing out, I question my ideas.
In, ideas.
Out, am I sure?
(Ten breaths)

Breathing in, I am still water reflecting.
Breathing out, I am aware of reality reflecting.
In, still water.
Out, reality reflecting.
(Ten breaths)

PART III
UNLEARNING WHAT IS KNOWN

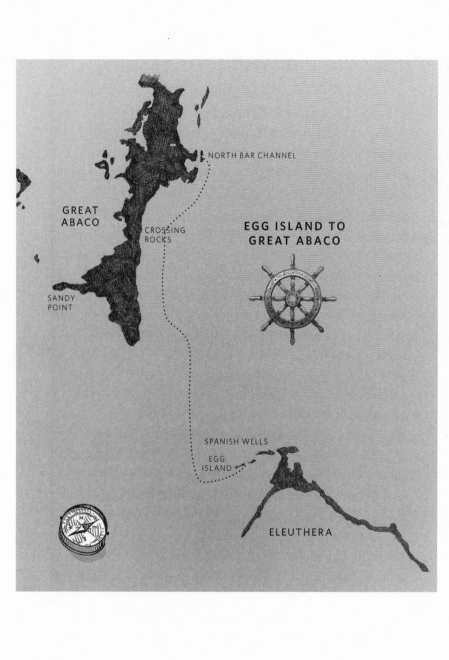

NORTH BAR CHANNEL

GREAT
ABACO

CROSSING
ROCKS

EGG ISLAND TO
GREAT ABACO

SANDY
POINT

SPANISH WELLS

EGG
ISLAND

ELEUTHERA

13 BEGINNER'S MIND

It's been thirteen months since I stumbled onto the beach in Barbuda. The week after we left, a friend from Wisconsin joined us in St. Martin, and for a while, wherever we sailed, we seemed to cross paths with cruising couples we had befriended along the way (to her, it appeared we knew everybody). Maggie and a friend of hers from college spent a week exploring Puerto Rico with us. Together, we did all the usual things tourists do (zip-lining in the treetops, visiting a rum factory), but we also pulled into Bahia Fosforescente for a unique treat: when night fell, the four of us climbed into the dinghy and whizzed around, dragging our hands through water to ignite the bioluminescence of countless aquatic creatures. The weird world shined green and gold.

I was seasick for nineteen hours crossing the infamous Mona Passage from Puerto Rico to the Dominican Republic; when we got to Hourglass Shoal in the middle of the night, Dave missed hitting a fishing boat by inches (it was driving an erratic course through a stormy sea). A few days later, when we were skirting the eastern shore of the D.R. (and again in the Mona Passage) I watched rain cells on the radar materialize from thin air over and over. I had no choice but to sail through the heart of the thunderstorm birthing center to get to the safe harbor on the island's north coast. The sea was tumultuous, and the boat pitched so violently that the spinning propeller in *Wild Hair*'s nether regions kept lifting out of the water, getting exposed in the troughs of the waves, making the engine's turbo charger grind hotter than an ax on a stone. But within twenty-four hours I was in Luperon at a beach barbeque, learning how to dance the Merengue from a local.

Wild Hair pushed through the chain of islands, back to the United States. The trade winds in the lower latitudes are supremely helpful to sailors pointing not southeast, as we did going down to the Caribbean, but northwest toward home. We stopped in the Turks and Caicos and the Bahamian islands of Mayaguana, Conception, Cat, Little San Salvador, and Eleuthera, all the way to Hope Town Harbor in the Abacos. Then, just before the start of hurricane season, we sailed two days straight from the Bahamas to Georgia (north of our insurance company's hurricane boundary). There, we crossed the finish line to safety.

This year, we skipped back to the Bahamas: Georgia, Florida, Nassau. We stopped off and fed scraps to the swimming, feral pigs who live on Big Major Spot before pulling into George Town Harbor. There, we met more friends from Wisconsin and took them on a weeklong cruise through some of our favorite stomping grounds. After they left, Dave and I started talking about the question I'd been dreading: when should we sell our beloved *Wild Hair* and become landlubbers again? We've done everything we set out to do, visited all the places we'd initially hoped to go. We've been thinking through our options, weighing the pros and cons of keeping her, but we haven't decided anything yet.

We're returning north now, visiting the Abaco islands again. I pay attention to the steady wind off the stern quarter making *Wild Hair* surf a spirited eight knots. If there is land around, it is past the horizon. I can make out only the sea springing and vaulting hither and thither. How I love this life without walls or windows, meetings or pantyhose, alarm clocks or snow and ice. The Great Atlantic Teacher has rehabilitated this advocate so completely that I hardly know myself as anything other than a sailor. The thought of returning to land makes me queasy. I'm afraid I may revert to the hyper-focused, stressed out, exhausted woman I once was.

There's a list of practical reasons for selling *Wild Hair*, however. Chief among them: the money we spent twenty years squirreling away was only enough to support a five-year voyage, and we've already eked out six.

There are emotional reasons, too. I just read Abraham Verghese's *Cutting for Stone*, and he says, "Home is where you're needed." That hit with the force of a tacking boom. Dave and I are needed on land more today than when we set sail. In 2007, Eland and Maggie were busy with college and my parents were feeling strong. Now we have a new daughter—Eland married his dazzling girlfriend from Mongolia. All of our kids graduated college and are bouncing across the country (and even Asia) chasing jobs and dreams. Maggie needs physical and logistical support. Eland and Khaliun are slogging through an immigration quagmire to establish her permanent residential status in the states. On top of that, my dad is in his eighties and my mom isn't far behind.

The wind gusts to twenty knots, so I turn off the autopilot, stand, and hand steer. My knees automatically bend to absorb the swells.

It's fitting that we're taking this route—sailing over the same ground we crossed on our first trip to the Bahamas, past Tilloo Cay where I experienced profound fear as *Wild Hair*'s engine broke and our anchors slipped while the storm pushed us down on jagged rocks. The memory of it makes me smile. That was ages and so many thousands of nautical miles ago. We've had very few mishaps in recent years. And no longer do other boats with more experienced crew sail past us; *Wild Hair* is the one who gains on the fleet and leaves ships in her wake. Keeping the boat happy and swift is as automatic as driving a car. Our knack for picking the right day and direction is proven. I'm proud of what we've accomplished (especially considering how little we knew when we started), but all my sailing skills will likely be irrelevant to life on land. I'll be a fish out of water, as they say.

"It's my turn to drive. Slide over, sweetie," Dave says, climbing into the cockpit with a soda and a fistful of pretzels. He's fresh from a morning nap. "You've been at it long enough."

"OK. The wind is at 1-6-0 blowing 20. You'll find the waves are a handful quartering the stern. You may want to take the boat in hand. Also, there's a ketch about two miles back at seven-o'clock." I put the ship on autopilot and move to the starboard bench.

"Got it."

We fall into silence. The boat moves beneath me and my waist and hips absorb the motion. Frothy peaks checker the steel-colored sea. Clouds—bulbous bags of water—tower overhead. But the sunlight streams unobstructed onto us; it glints from all the shiny surfaces on the boat and ocean and I fantasize I'm at a fairy convention and there's magic in the air.

The motion lulls me and my focus softens over the port rail. Maybe I should take a nap myself, I think. Suddenly, a hundred yards off, the water explodes as a massive whale breaches and tips. It strikes the water and spray blasts outward in horizontal arcs.

"AY, AH, AYEEE!" I shout, pointing. "What?" Dave exclaims in panic, looking. "Humpback—huge!" I spew.

"Where?"

"It breached there." Both of us stare west.

"Where is it?"

"It's gone, but keep looking."

It was a fifty-foot giant that leapt halfway from the water and was perfectly lit by daylight. I saw the animal's pointed head (which looked flattened by an anvil); long white flippers, scalloped along trailing edges; and near-black body. The whale busted free from the depths and made its presence known. When it did, I heard a clear and commanding voice say: "I'M HERE. WHAT ARE YOU GOING TO DO ABOUT IT?"

"It was gorgeous . . . bursting," I stammer. "It leapt and slammed sideways like—like it wanted to see the sky and make a huge splash just for the fun of it."

"Huh, we haven't seen humpbacks before, even though we've been sailing their migration route all this time."

"No, we just saw those hundreds of pilot whales with their short and droopy dorsal fins that one time near Andros."

"You saw them. I was napping." I look at my man. His disappointment registers.

"Well, that looked like a huge pod of unusually long dolphins. They weren't very whale-like. Honey, I'll bet where there's one humpback, there's more." I step over the cockpit and onto the deck for a wider view. After a moment, a mini volcano of water blows skyward from the surface, then another.

"There—can you see them breathing? Watch for puffs." Another whale exhales closer in.

"I see it," Dave says, turning on the autopilot and stepping on deck. The sun is behind us as we scan. Then—only ten feet from *Wild Hair*'s hull—a colossal humpback mother and baby roll near the surface. One expressionless eye from each takes us in. The pair is attentive and appraising.

"Mother and baby—right there!" I whisper standing frozen.

"I see, I see." We fall mute. The pair glides, their tails barely move. Or maybe they're still and it's *Wild Hair* that's in motion, I don't know. The mother is as long as our ship. Her body is habitat to colonies of barnacles and her skin is mottled with bright splotches, like clumps of snow on asphalt. The baby is a fifth her size, with smooth skin in a lighter shade of gray. The youngster's tail appears sharp-edged and muscled. They roll their bodies further over to maintain focus. Unconsciously, Dave and I walk alongside our mammal cousins down the length of the deck. But too soon we're out of room and our friends disappear into the near black water behind the ship. It's then I see the other sailboat traveling behind us between Eleuthera and Abaco. Handhold to handhold, I work my way to the cockpit radio.

"Ketch sailing the Northeast Providence Channel to Abaco, Ketch sailing the Northeast Providence Channel, Ketch sailing the Northeast

Providence Channel, this is sailing vessel *Wild Hair*, *Wild Hair* to your north. Over."

"*Wild Hair*, *Wild Hair*, this is *Mister Mustard*. Over."

"*Mister Mustard*, there are humpback whales in these waters. They are breaching. A mother and baby passed close to our ship and they are headed straight to your bow. You may want to keep an eye out. Over." At a distance, I see a woman on *Mister Mustard* step from the cockpit to the deck; she's scanning.

"*Wild Hair*, *Wild Hair*," the man's voice continues, "this is *Mister Mustard*. Thanks for the very exciting news. We'll be sure and keep a look out. *Mister Mustard*, over."

Dave and I continue to watch the surface, but there are no more whales nearby. Our excitement fades, leaving me bluer than ever over the thought of selling *Wild Hair*, our vehicle for adventure. There's so much out here waiting to be discovered.

I come alive in the company of the Great Atlantic Teacher. How will I stay connected to nature's rhythm when I live in a house that refuses to move with the wind? Who will I be when I can do what I want, when I want, and ignore not only the conditions of the day, but the cycle of seasons too? What actions will I take when life no longer needs my vigilant attention to safety? This person who will carry a purse and wear boots and coats, my future self, will I even like her?

Two hours pass and the wind is blowing hard and constant now. The ride is getting unruly. *Wild Hair* moves well, but my body is growing fatigued. I'm ready to call it a day and settle down to pasta, a salad, and a good book. *Mister Mustard* took another route, and we are alone a mile outside the North Bar Channel. With the engine running, Dave and I roll in the genoa and drop the mainsail. Then Dave spins the wheel and points our ship toward the cut while I tidy lines and secure water bottles, winch handles, and other loose equipment for the last leg of the trip. We will follow the range markers between Lynyard and Pelican Cays into the still waters of the inland Sea of Abaco. Soon, we'll anchor in Marsh

Harbour—a stone's throw from where we spent Christmas with the family on our first Bahamian visit. Maybe, I think, we should start watching the new season of *Downton Abbey* tonight.

"Heather, can you see the entrance?"

"What?" I look toward the cut. "What do you mean?"

"The chart says it's straight ahead, but I can't see it." Standing on deck, I make out slivers of land above the ocean, but the normally razor-sharp land-water edge is fuzzy. I blink and stare. I think my contacts are blurry in the wind. After a moment, the water appears to drop and a clear view of the cut opens.

"We're pointed straight at it. It's there, I see it."

"Huh." Dave's brow twists. The furling line controlling the forward sail snakes in a mess around the aft winch. Because of the way *Wild Hair* is bounding, it's impossible for me to stand and neaten it up with two hands, so I kneel low on the deck and pick up the line. I do what I always do: challenge myself to make two-foot loops (each one identical to the next) because I want the bundle to look ship-shape when stowed in plain sight.

My back is tired. We left at sunrise and rode this bucking animal for eleven hours. That's more than enough for one day. In just a few minutes, on the other side of the cut, *Wild Hair* will be in the placid waters of the Sea of Abaco. Everything will lay down flat in the lagoon behind the barrier islands.

Nearer shore, waves start to roar. That's odd. I lift my head and notice a curl of ocean unpeel from a quarter mile off toward me. The breaking wave is taller than the boat. This isn't right. I've never seen these conditions. I stop what I'm doing and move inside the cockpit to put on my life jacket. I hand Dave his jacket without saying a word.

"Good idea," he says. I sit next to him at the helm to get a view of the electronics and help monitor what's happening. Things are intense. The giant limpid curls (which the sun illuminates in beautiful hues of turquoise) are deafening and savage. The power of their energy is brutal and it dawns on me that these monsters are violent enough to destroy the ship.

"I don't want to be here," Dave says. "Baby, this is a 'rage sea'—the very thing every boater in the Bahamas knows to avoid. We shouldn't be taking this cut today."

Holy shit, I think. He's right. All the guidebooks warn about the Abaco Rage. Crazy seas happen here because beneath the water there is a seam where the ocean bottom meets the Bahama Bank, and the sea floor rises 650 feet in just a mile. When strong easterly winds blow for a period of time, they drive ocean swells west, and the abrupt shallowness creates wave masses the size of multi-story buildings. The rage sea can make this passage between the barrier islands impossible to navigate. And here we are.

"No, we shouldn't be here." I swallow hard.

"I can't turn around. If I turn the boat sideways, the next wave will make us roll and break apart." Dave's tone is three-fourths statement and one-fourth question.

"Yes, that's right. Don't turn."

"So, we're committed. We have no choice but to go through the cut."

Suddenly, I remember the plight of the forty-six-year-old woman— Laura Zekoll—who crewed in 2010 on the ill-fated Jeanneau 45 named *Rule 62*. That boat, which was like *Wild Hair* in more ways than not, was part of the Caribbean 1500 Rally from Virginia to Saint Thomas. They were just eight days ahead of Dave and me the year we sailed to St. Thomas from Jacksonville, Florida. Because of seasickness, *Rule 62* diverted to the Bahamas and tried to enter the Sea of Abaco through the North Bar Cut—this very channel. They arrived at night during a rage sea. *Rule 62* was lost. Three crewmembers survived, but Laura was never found. Once, I saw a photo of *Rule 62*'s route online (all the boats in the rally had GPS trackers onboard). To me it looked as if the crew tried but failed to turn the ship around at about where we are now—the beacon traveled in a sharp left-hand turn and then stopped on the rocky shore just there.

"Keep going straight. Don't turn."

I also remember hearing that in 2011, a rage sea put the local mail delivery boat—a 160-foot motor vessel named *Legacy*—on the same rocks off our port rail. How did the locals who travel these waters every day get into so much trouble? That time, the steel ship didn't break apart, but a US Coast Guard helicopter needed to come over from Andros Island to rescue the crew just the same.

"I can't tell where the center of the channel is." Dave's voice is strained high. "The chart says the cut is twenty-two feet deep and the electronics says I'm in the middle, but the depth meter says I'm in twelve feet of water. What do I do? Go right or left?"

"Feel your way, baby. Try motoring a little in one direction then the other." Dave tilts the wheel left and the depth quickly drops to eight feet. "No—try the other way." The depth increases to nine and then drops to seven. "Baby, you may be there." To help Dave keep his eyes on the boat and the environment, I start calling the depth readings out loud. "6.9 feet, 6.4, 6.4." We both know the keel will strike bottom when the depth reaches five feet.

To the sides and behind *Wild Hair*, waves lift high overhead and break. If our keel hits bottom and we come to a sudden stop, *Wild Hair*'s momentum will tip the ship onto the bow and the curling wave sucking the water up from under the hull will lift the ship's back end. In an instant, *Wild Hair* could summersault—ass over tea kettle. Toppling into a nosedive will buckle and snap *Wild Hair*'s rigging, shatter the hull, and toss Dave and me from the wreckage. There's a good chance our bodies will never be seen again.

"6.3 feet, 6.1, 5.9 feet." I grab hold of the stainless post supporting the canvas bimini overhead. Around us, the sound is something of a cross between a train and a jet aircraft. Dave has his entire being focused on keeping the boat straight as the wave energy violently yanks us and tries to make *Wild Hair*'s tail wag.

"5.4 feet of depth—where did the channel go?"

"I don't know!" Dave shouts. "The chart plotter says we're down the middle."

"5.3 . . . 5.3 . . . 5.2 . . . 5 FEET," I shout. My body seizes for impact. "5 FEET, BABY. 4.9 FEET . . . 4.7 FEET . . . 4.5 FEET."

This is it, I think silently. A weird calmness washes me. This is how it happens. Now, we're dead.

And yet, impossibly, *Wild Hair* keeps moving and Dave keeps steering. We can't motor through water more shallow than our keel. What's going on? I quit reading the depth gage and look around. To my right, a peak builds higher than I imagined it could. All the water in front, underneath, and to the side of the ship rises into the tower in a great ascension. When it reaches more than two stories overhead, the peak leans toward the coast and curls into a tube. The interior of the tube glows and opens (in a fraction of a second) into a tunnel that runs the length of the coast. Then, the energy of lifting and rolling transforms and white froth pinches the tube nearest to me shut. Water collapses, filling in the hollow.

The depth meter continues to read 4.5 feet. *Wild Hair* will not survive the soon-to-be tumble. We will not survive. I breathe, completely undone, and put my eyes on the depth gage again. It reads: 4.5 FEET . . . 4.6 FEET . . . 4.8 . . . 5 FEET.

"5.2 FEET, Dave, we're at 5.3 FEET!" He says nothing in his concentration. I continue to announce the numbers. "5.7 feet . . . 5.9 feet . . . 6.3 feet . . . 7 feet—Honey, we're at 7 feet!" The sound of the roaring ocean is at a slight distance now. We are coming to the far side of the channel, almost to the Sea of Abaco. The surface is lying down. "8.4 feet . . . 9.9 . . . 12.3 feet. DAVE, WE MADE IT THROUGH. We made it through." I slump onto the backrest, panting.

I look at my man, who relaxes only enough to turn toward me. He's shell-shocked, the color of parchment. His pupils are pinholes. Dave lifts his right hand from the wheel and extends it my way. His fingers are trembling. I lift my arm and see my hand quakes, too.

"Where the fuck was the channel?" Dave whispers. He is spent, but swings *Wild Hair*'s rudder ninety degrees to follow the course north.

Indeed, the Sea of Abaco bears no resemblance to the entrance we just traversed. It's serene. It's surreal. The far-off roar is dim.

"It looked like the giant waves were made from channel water," I say. "I could see the towers building from the ocean around us. They sucked everything in, so we lost our depth to the peaks, I think."

"Oh, my god, honey. That was the stupidest thing we've ever done. We never should have entered the range when we couldn't see it."

"You're right. I did see it, but only for a second. That wasn't enough."

"And I never did. We should have stayed offshore and circled until we came up with a different plan. I can't believe we just survived a rage sea. That's what sinks ships and kills people every year."

"I thought we were dead."

"We're lucky to be alive."

I hold my hand out and watch my fingers tremble again. Dave slides off his life jacket, but I'm not ready to part with mine yet.

"You've got to promise me something," he says. "We can *never* tell anyone what we just did. It was too stupid, and every single sailor knows better. We'll be a laughing stock."

"You're right. I won't tell anyone. We're better sailors than this. It never happened." I stand and swing my arms to shake off tension and think about the harsh commentary bloggers hurled at the captain of *Rule 62* after Zekoll lost her life. Sailors, I think, can't imagine themselves making a catastrophic mistake or going blindly into circumstances for which they are unprepared. If it truly could happen to anyone, they'd get spooked. They need to believe they're immune to their own stupidity. I know because up until this moment I took comfort in my expertise, too. The North Bar Channel shattered my bravado and humbled me.

"How long do you think we were in hell?" I ask.

"No more than five or six minutes." I nod in agreement.

"It's so weird; it's beautiful in here and the sailing was great all day."

Dinghy's little face appears at the companionway screen. "Hello, you! You slept through all the action. Wouldn't you have been shocked if we rolled and you went for a swim." I clip her collar to the leash at the helm and she jumps onto the seat. A seagull swoops overhead, making Dinghy duck and then crouch low to pounce should the bird return. Her butt wiggles as her back claws dig into the upholstery for traction.

"The thing I don't understand is why the sea is raging today. I was thinking rage seas happen only in stormy conditions."

"No. I guess all you need is a strong wind from the East or Southeast for a few days and that mounds the ocean. We certainly didn't have those conditions the other times we went through." He looks at me again. "God, that was awful." I breathe and look at him. I have no words.

"OK—so how did *Wild Hair* sail through water only four-and-a-half feet deep?"

"I haven't a clue," Dave says. "We know the depth gage is accurate. I cross-referenced it with the other electronics and measured it against the length of my body when I dove on the hull in George Town."

"Do you think there was so much turmoil that the sand suspended in the water above the ocean floor confused the sensor?"

"That makes sense." Dave takes his windbreaker off. "Whatever happened, we know the reading was wrong. This boat in that channel would have pitch-poled at five feet." In unison, our minds return to the horror of what almost was.

"I thought we were dead. That's why I stopped reading the depths out loud."

"I thought we were dead, too."

A long silence settles as fear ebbs and exhaustion floods our bodies. My breathing normalizes as the radio comes alive with other people's happy voices making dinner reservations and plans to go scuba diving in the morning. I can't fully relax, however, because a question needles my consciousness. After thirty years of studying sailing, owning numerous boats, and six years of living aboard *Wild Hair* and sailing to Grenada and back, how could we have been so stupid to enter the North Bar Channel during an Abaco Rage?

I would have been less surprised had this happened on our first visit here, or our second. We had a lot to learn early on, lessons that can only be ascertained by trial and error. But we paid attention, experimented, and then finally caught on to the nuances and intricacies of safe sailing in the ocean wilderness. We spent years practicing our skills day and night, sharpening our know-how through an ever-increasing range of conditions. Eventually, Dave and I developed a real expertise in taking care of this boat, knowing when to sail, and getting to our destination as planned and without calamity. For years we traveled over the horizon without any major self-inflicted sailing gaffs.

But today, at the moment when we know the absolute most, on a boat fully equipped with advanced technology and in top condition, when our bodies are strong and our minds are attentive, the two of us sailed headlong into the most dangerous moment we've experienced aboard *Wild Hair*. How is it that we, two seasoned sailors navigating a familiar course on an otherwise lovely day, nearly killed ourselves? Are we uniquely senseless? Or does humankind have a natural propensity, not toward wisdom, but stupidity? I contemplate this question for many minutes before deciding my species has the capacity for both wisdom and foolishness in any given moment. We can either be profoundly insightful or completely forgetful depending on our degree of awareness.

Now that I think about it, Dave and I would have been way too afraid of the North Bar Channel's reputation to even take this route when we first started sailing. In time, cautiously, we did venture through it. Several times even. This time, I assumed I knew what the channel was like. Experience made me brave, and I didn't take the time to empty my mind of what I thought I knew to make room for the truth of *this* moment, *today's* conditions. I approached the cut like an expert, entertaining only a few known possibilities. I didn't approach the pass like a beginner with a mind open to the full range of possibilities under the stars.

Why was I so forgetful? I was tired and focused on being done.

My decision-making was mechanical, mindless, and based on my past experiences.

Things would have unfolded differently had I approached the channel with what Thich Nhat Hanh calls "beginner's mind." My thoughts would not have leapt ahead to tonight's meal and entertainment choices. I would have reflected less about what I already knew and more about what I didn't know. I would have noticed earlier that the cut was in fact behaving in a way altogether different from anything I'd experienced and said, "Hey, Heather, why *does* the shoreline look like jumping elephants?" Had I a beginner's mind, I would have approached the North Bar Channel not on autopilot but with my awareness dialed into the reality of the danger ahead.

Giving up the been-there-done-that fallacy of already knowing what there is to know, opening my consciousness to unknowingness and growing curious about the details of now, preconceived notions fade and the present comes alive. This is how to practice beginner's mind: focus not on what's known but on what is unknown. Doing that, I can meet events freshly and stop clinging to the misguided notion that this moment is like the past. Attentive and ready for anything, I can be as creative as an artist in improvising new and effective ways of being appropriate for *this* circumstance.

Beginner's mind is the most precious form of mindfulness, because unknowingness is the most intimate wisdom.

As Confucius said, "To know that you don't know is the beginning of knowing."

Finally, I take off my life jacket and tuck in my t-shirt. It occurs to me that maybe the Great Atlantic Teacher just delivered her most powerful lesson: I should return to my life on land with beginner's mind. Rather than believing in only a limited set of possibilities for myself, I ought to empty my mind of both preconceptions and worries. I should admit I don't know what the future holds. I should better cultivate mindfulness and be curious about the reality of now, how it unfolds. If I come from *unknowingness* on land, I can be warrior-strong when confronting trials and as creative as Michelangelo himself when sculpting a new life from the raw material of possibilities. Of course, I will slip into forgetfulness from time to time, but it will be hard to forget altogether this instruction from the Great Atlantic Teacher: cultivate and maintain a beginner's mind, or else.

REFLECTIONS ON BEGINNER'S MIND

In July 1992, at age thirty-two, I was at the helm of a rented twenty-nine-foot sloop and sailing back to port on the last day of a family vacation. The four of us had really taken to our first nautical adventure. Eland and Maggie, who were five and three, woke Dave and me every morning at about 06:30. I'd make breakfast in the doll-house-sized galley while they dressed and tidied the boat for sail. Then, we'd set off in a new direction, toward some anchorage in the network of land masses dotting Lake Superior's Apostle Island National Lakeshore. They were quick sails and Dave did his best underway to hook Jumbo Lake Trout or Coho Salmon. By lunchtime, we'd drop anchor and step ashore for new discoveries and adventures.

We stomped through virgin forests, constructed sandcastles on the beach, spied eagles, and climbed the spiral steps of lighthouses. Eland never removed his plastic cowboy hat and western-style gun belt (loaded not with a toy gun but with stones and pinecones). He yakked on without pause about dangerous characters he'd made up lurking behind the trees. Attentive to his every word, Maggie kept pace in her snake-skinned cowgirl boots. Only occasionally did her concentration swerve toward an entertaining bug or a dead fish on the beach. One afternoon, after strapping on a life jacket and tying himself to a long line (so we could retrieve his body if necessary) Dave whooped and hollered as he tried to swim in the 53°F blue-black water.

The whole time we were sailing I was leery of Lake Superior. At 350 miles long and 1,300 feet deep, it's the largest lake in the world and holds ten percent of Earth's fresh water. Lake Superior is big enough to create weather, and it happens unpredictably. Sudden shifts have sunk hundreds of vessels throughout the centuries. Once sunk, the cold water stops microbial growths that would otherwise produce gas and float human bodies to the surface. That's how Lake Superior earned its reputation of swallowing ships and mariners forever into its depths.

Nervous about the lake's reputation, I found comfort in the good boat we leased. She was a beefy, solid feeling monohull that performed well the entire week. I had taken a how-to-sail course the previous spring, and I got the boat to go from place to place all week without too much hubbub.

But then, as I skippered the ship back to the marina to return the boat and drive home, the wind and waves kicked up. Gusts pushed hard against the mainsail and caused us to tilt more than I liked, before backing off so we could bob vertical again. With each blast, I let go of the line a bit to take tension out of the mainsail. I turned the bow into the wind. These actions spilled the force and kept us from blowing over, but I really had my hands full managing the bursts.

Dave sat near me in the cockpit eying the shore. Dense forests of deep evergreen graced hilltops 1,000 feet above. This US National Park was long ago tagged the "Caribbean of the North" and the name fits. Twenty-one islands scatter like seeds to the four winds, and each landmass sits about two miles apart. This is one of the world's best sailing grounds.

Our little boy, dressed in a life jacket and tethered to the boat, was asleep and drooling on the cockpit floor. He had loaded his gun holster with tiny cars and one had fallen out next to his elbow. Maggie napped in the forward V-berth below decks; her boots stood next to the bed on teak floorboards. The rhythm of three-foot seas lulled everyone but me.

Suddenly, an especially savage blow filled the mainsail and this time it didn't quit. The boat tipped over quite a bit to the port side.

"Holy crap," I shouted as I released the sail's lines and turned the boat toward the wind. But unbeknownst to me, a twist in the rope (a kink I would later learn sailors called an "asshole") kept the sail bound in place so the canvas couldn't swing loose. The boat leaned further askew and the starboard hull crept out of the water to reveal the slippery skin of the ship's underbelly, seen previously only by trout.

"Oh, my God," Dave blurted as he and I climbed to the starboard edge, added our weight to the high side, and tried to make our bodies ballast to right the ship. But the gust kept coming. Our incline increased.

Eland screamed. He had rolled to the down-side of the cockpit and the flood of chilly water overtook the port rail and sluiced him head to toe. The dousing carried his light frame to the back of the cockpit and only his tether kept him on board.

I stared in horror at the boat's slow-motion fall, as the angle between the mast and the water closed to nothing. The next wave spilled over the mast and the entirety of the mainsail slipped under a shroud of Lake Superior waves in one gulp. Dave and I lay our bodies as far out across the hull as we dared, but we were ineffectual counterweights.

After what felt like eons, the brutal air current let up. Water drained from the scoop of sailcloth and the mast lifted above the surface. The keel kept levering the stick (slowly, slowly like a minute hand creeping toward noon) to its upright position. Dave spotted and released the twist in the line and the bow swung into the wind, so the feral sail flapped sheets of Lake Superior on top of our heads.

"Would you take over?" I asked Dave as I scooped a crying Eland into my arms.

"Yup," he said. I unhooked Eland's tether, and brought him below for some comfort, warmth, and a change of clothes. Our girl was still sleeping but her body and boots had tumbled to port.

The experience completely freaked me out. I was done sailing the boat for that day. As a matter of fact, I was so frightened I refused to even step *onto* a sailboat for the next ten years. I only got interested in sailing again after someone at a dinner party happened to understand and explain to me the mechanics of what occurred. I learned we experienced what's called a *knock down* and it could have been prevented had I positioned the sails not tight for the soft blow but loose in anticipation of the gusts. Also, my dinner companion described for me the look of a gust as it comes across the water, the way it shreds waves in a dull gray line as it nears.

I needed to cultivate beginner's mind before I could step back aboard a sailboat again. I had to let go of my very strong sense of being helpless and afraid (a narrow perspective based on limited experience) in order to see that the reality of sailing held a broad range of possibilities for me. Sailing, in fact, contained six years of exploration, wonder, awe, insight, and just plain *fun*. It was beginner's mind that launched me on this physical and spiritual voyage through the Western Atlantic and Caribbean Sea, and it gave me courage enough to discover clues from the Great Atlantic Teacher on a new way of being in the world. What's more, it is beginner's mind that will carry me back to my life on land again so that I might learn and do . . . who knows what. And now that I think about it, beginner's mind can completely transform how I approach the climate challenge.

I need to stop thinking the world is like it was when I was growing up. Deep down, I know it might look similar, but the present isn't behaving like the past (and the future won't be like the present, either). The National Aeronautics and Space Administration (NASA) and the National Oceanic and Atmospheric Administration (NOAA) tell us the climate is changing, humans are responsible, and humans and natural systems are at risk.[43] Nearly all independent scientific agencies at home and abroad agree, and what's *causing* the warming (according to study after study) is a suite of gases: carbon dioxide, methane, nitrous oxide, fluorinated gases, and a couple others.[44] The most important thing to know is *human activities*—specifically our transportation industry, electricity and heat, and the burning of forests—are putting unprecedented amounts of carbon dioxide into the atmosphere, and animal agriculture is putting a dangerous amount of methane into the air too (because a lot more people today want a lot more meat).[45] Counting our chickens (and cows, sheep, pigs, and all the other critters we find delicious), there are now more livestock living on the planet than people, and there are 2.5 times as many people living on Earth today then the day I was born.[46] Do you see what I mean? Things are very different now, and not in a good way.

It's impossible for us to carry on as our parents and grandparents did and expect the same results. NASA reported that February 2016 was 1.35°C above normal, making it the "most unusually warm month" measured in human history (the month before, January 2016, had held the record previously).[47] What appears to be a benign warming of a degree or two completely upsets a delicately balanced global apple cart. For the next 25 years or so, scientists expect conditions to seesaw between hotter and colder, wetter and drier, and for there to be fewer but more intense storms. Low-lying coastal and island states, the vast deltas of Asia (including India), the planet's Polar Regions, and most of the continent of Africa will be most vulnerable to the changing weather patterns, as these places are highly dependent on subsistence agriculture. Catastrophes will strike these regions, scientists say, in the next decade or two.

In the mix of climate concerns there lies a gross injustice. Pope Francis, in his encyclical letter *Laudato Si',* speaks to a long history of wealthy nations consuming the resources of developing countries in exchange for ruin.[48] Under and nonregulated international practices of extraction, coupled with short-term boom-bust economics, have repeatedly destroyed vulnerable people's resource bases, health, and social fabric and left behind a legacy of toxic pollution. Pope Francis (who spent countless hours conferring with scientists) urges developed nations to pay their debt to the poor, first by consuming less fossil fuels, and second by helping struggling nations transition to green energy and sustainable development practices. Looking deeply into climate change with beginner's mind, the pope envisions fresh possibilities for fairness, justice, and equality.

Sometime after the next twenty-five years (we don't know exactly how soon), if we continue behaving as if nothing has changed, the weather will become unpredictable enough to wash away the world's coastal cities, cause mass extinctions and food shortages, and potentially spark wars.[49] It's no wonder sixteen top US Generals and Admirals recently declared climate change to be a "catalyst for conflict" and a threat to the US economy.[50]

Many think the root of the climate challenge is the human population explosion in the last one hundred years. Yes, our numbers are clearly straining planetary systems (especially since our behaviors haven't shifted to compensate for our numbers). But one team of scientists recently developed a tool to help us define and collectively stay within Earth's safe operating space: they call it the Planetary Boundary Framework.[51] The group identified nine disrupted global systems on which our lives depend, including: climate change, ocean acidification, stratospheric ozone depletion, nitrogen and phosphorus cycles, freshwater use, land use, biodiversity loss, atmospheric aerosols, and the introduction of something they call "novel entities" (like pollutants, radioactive materials, and micro-plastics). Researchers say if we can *steward* these nine systems, then life on Earth has a very good chance of continuing and flourishing.

We have it in us to pay attention and look around with beginner's mind at the full range of possibilities before us. When we finally look in wonder, awe, and insight at the world of possibilities, we may give rise to a future we prefer, one that meets the needs of our global human and nonhuman community.

It's too late to steer clear of all of the effects of climate change. But there's a lot we can do as individuals to turn things around, and *small actions done by many people make a big difference*. We can, for example, cut way back on our meat consumption, especially beef (to minimize the quantity of methane released, and protect forests from being burned for their pasture). We can drive fuel-efficient cars, walk, ride a bike, or take public transit. Fewer airplane trips each year can make a big difference too (or we can at least compensate for our airplane emissions by buying carbon offsets through the airlines or through the marketplace with the help of companies like NativeEnergy.com and TerraPass.com). We can reinsulate our homes, install a smart thermostat, and let rooms be a little warmer in the summer and cooler in the winter. We can buy local to cut back on shipping, plant trees, grow our own vegetables, air-dry clothes, reduce, reuse, and recycle.

We can tune into political conversations and vote for candidates who share our concern for Earth and her children. We can speak out in support of state and national policies that favor green and renewable energy over fossil fuel. We can write letters to elected officials or the editor of local newspapers, participate in local marches, and talk with friends and family members, urging them to do the same. When we start to take action on the climate challenge while coming from a place of beginner's mind, we discover ourselves in the company of millions of everyday heroes who are already doing their best to sail a route straight and sure.

Scientists tell us time is of the essence. The sooner we clear our heads of the individual and collective plans we made in our youth (those shaped by a now dangerously outmoded set of assumptions), and get curious about the truth of Earth's new reality, the better off we and the other species on Earth will be. It's time to stop thinking the future will be like the past, and start cultivating enough humility and beginner's mind to rethink everything we thought we wanted in our lifestyle and our priorities. We are, after all, entering a strange new world for the very first time and we have no expertise in navigating the turmoil ahead.

When we practice beginner's mind, we consciously stop leaping ahead to thoughts of an idealized, unrealistic future or images of our own demise based on memories of humankind's *worst* moments in history. Similarly, we open to fresh possibilities when we set aside our preconceived notions about politics and business. We let go of these things because beginner's mind is first and foremost about our *direct experience*, and this is shaped by and this includes time in the outdoors, our own reading and exploration of scientific discoveries, calm conversation with others equally determined to protect life, and an ecological mindfulness practice on and off the cushion to cultivate inner knowing and insight. Beginner's mind will naturally lead us to reflect upon our own daily actions and personal lifestyle choices. It may also inspire us to join with others in larger nonviolent actions aimed at environmental conscious-ness-raising and societal transformation.

This moment is unlike any that has come before in human history or will come again. When we know this, every action we take in daily living becomes a strategic choice to protect ourselves and Earth. Cultivating beginner's mind, we wake ourselves up to the potential of now, rise like spiritual warriors on the battlefields of our own consciousness and meet the artist within—our own spirit, with the creative capacity to respond freshly and skillfully to what *is* here *now*. The personal expertise and know-how we have amassed in our lifetime will be useful; we can con-tribute our skillsets to help navigate a healing path.

Beginner's mind is not a fixed state of being, but rather a practice of looking and seeing deeply, again and again. To train myself to look with fresh eyes, I wrote this meditation. I sit with this meditation at least once every two weeks.

Breathing in, I see the world of my childhood
   free of climate concern.
Breathing out, I smile to the plans I made for myself.
In, free of climate concern.
Out, my plans for myself.
(Ten breaths)

Breathing in, I see all the ways I thought the
   future would continue like the past.
Breathing out, I let go of my misperception.

In, see my misperception of the past continuing.

Out, let go of misperception.

(Ten breaths)

Breathing in, I grow curious to discover all that I
    don't know happening in this moment.

Breathing out, I know what I don't know is the beginning of knowing.

In, curious to discover what I don't know.

Out, beginning to really know.

(Ten breaths)

Breathing in, I see many opportunities to take
    care of myself and Earth here and now.

Breathing out, I see a warrior inside who can act with skill.

In, many opportunities to act.

Out, awaken my inner warrior.

(Ten breaths)

Breathing in, I see many opportunities to take
    care of myself and Earth here and now.

Breathing out, I see an artist inside who can respond with creativity.

In, many opportunities to act.

Out, awaken my inner artist.

(Ten breaths)

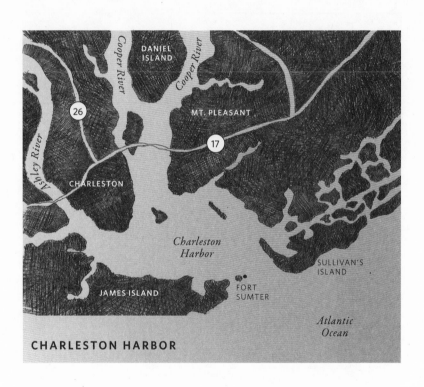

CHARLESTON HARBOR

EPILOGUE

I'm wearing an Operational Dress Uniform and standing in Boeing's parking lot in Charleston, South Carolina. It's the airplane giant's "Family Day," and spouses and children are buzzing to enter the restricted warehouses and see full-sized wings and tails strewn with the organized chaos of Santa's workshop. Out here, the makeshift canopy provides little relief from the scorching sun. (I'll wager the radiant heat off this asphalt puts my micro-climate at nearly 100°F.) Military boots, thick trousers, a shirt, jacket, and cap are way too warm for this urban heat island.

"Flotilla 12-8 is a great group. It includes people of all ages. You should think about joining." I say to a visitor, a retired soldier. I'm talking about the benefits of membership in the US Coast Guard Auxiliary.

"Maybe I will," she says. Her eyes scan the handouts smothering the table.

"You can teach boater-safety classes, be a vessel examiner, or—what I like—volunteer as a radio watch-stander to monitor Coast Guard patrols and listen for recreational and commercial vessels in distress. You know—Maydays?" She nods, but I know the woman can't imagine my history, what drives me to help US Coast Guard rescue operations, how determined I am to pay forward the kindnesses extended to me. She shifts her weight as her two friends stand patiently, sucking the straws of super-sized sodas. I wipe a bead of sweat from my temple, catching it before it reaches my jaw.

"Auxiliarists train as crew on Coast Guard vessels, too, so we can ride in those awesome boats, help out on patrols, contain toxic spills in the ocean, and whatnot. Some Auxiliarists even train for service in helicopters."

"Man, I gotta do something. Just getting back into a uniform would be a relief. Then I could stop having to worry about what to wear!"

I laugh. "We meet monthly. This brochure has the details. Come check us out." The woman agrees and disappears with her friends into the crowd. I look past the table at Dave, who is handing out coloring books and talking with parents about life jacket requirements and the flotilla's next safety course. He's aged from blonde to silver, but somehow looks even more adorable in a uniform.

We sold *Wild Hair* during a North American cold snap that was so severe people named it the first "Polar Vortex." After spending six winters in the islands, Wisconsin's record low temperatures shocked my system. Snow that fell in November didn't melt until *June*. I found myself house-bound, depressed, sedentary, and fat. I positively ached for the company of the ocean and the sun's warmth. So within a year of *Wild Hair*'s sale, Dave, Dinghy, and I (along with our new Pomeranian puppy named Dharma) packed up and left Madison—our home of thirty years.

We settled outside Charleston, South Carolina, in a house with a big front porch and a classic porch swing. Loblolly pine, magnolias, camellias, and live oaks cluttered with Spanish moss grace my yard. Eastern tiger swallowtail butterflies flit around flowers. Carolina wrens, chickadees, house finches, American goldfinches, brown thrashers, ground doves, northern mockingbirds, cardinals, brown-headed cowbirds, eastern bluebirds, red-breasted nuthatches, and tufted titmouses come to my feeders. Killdeer skip through the yard, and pileated woodpeckers hammer softly on the bark of my palm tree. Ruby-throated hummingbirds circle sugar-water troughs just outside the window where I write.

American robins stop by too, but mum's the word these days. I love their company nonetheless.

Dinghy passes her time stretched on the windowsill, chattering at the particularly plump cardinals who come near to dine on sunflower seeds. Every chance she gets, she darts past our legs through open doors hoping to go for a walk-about. (She doesn't appreciate her demotion to house cat). My parents bought a home three blocks away from us, and made the big move from Arizona. Eland and his wife, Khaliun, also picked up and left Mongolia to live in a nearby flat, start careers in the US, and grow a family. The pair is in publishing: Khaliun was editor in chief for *Cosmopolitan Magazine* in Mongolia and now manages from afar the only online lifestyle magazine in her native land; Eland (after stints at *National Geographic Traveler* and *Mongolian Economy*) became editorial manager for a local publishing house in Charleston. Maggie, the outlier in the bunch, lives in Kansas City with her boyfriend. Our animal-loving girl grew up to be a penguin and polar bear zookeeper.

I am a northerner transplanted to the Deep South, so I continue to feel like a traveler, a newcomer in a strange land rich with a history I barely understand. A little research revealed that in 1630, the soil on which I dwell was a rice plantation. The landowner, in 1730, switched to growing cotton. Only a month after Dave and I moved in, a violent crime of racial hatred at Mother Emanuel AME Church killed nine worshippers in the midst of prayer. Humanity's long struggle for justice is in the air and sometimes, when I'm meditating in my garden next to Egyptian papyrus and under climbing roses, wisteria, and confederate jasmine, cries from generations who suffered and died then and now echo in my body. I breathe in pain and breathe out compassion.

Water is a constant presence. Local rivers, harbors, and the ocean can't seem to decide if the territory is solid or liquid. I love the formlessness of Nature's indecision here, but the water teases and tantalizes me and not having a boat makes me restless. Going to the beach doesn't help—my eyes leap to the horizon and my body feels pulled to go *out*.

Again and again I grow mindful of my craving to be other than land bound and let the sensation pass. Dave and I *are* casually sniffing about for our next boat. (And Dinghy is all for the idea.) But *Wild Hair* is a nearly impossible act to follow and we have yet to meet the right one.

And then, of course, there's Charleston. According to legend, I've moved to the birthplace of the Atlantic. The ocean was born in Charleston Harbor at the confluence of the Ashley and Cooper Rivers. Charleston is nicknamed the "Holy City" because its skyline zigzags with church steeples. There are more than 400 places of worship serving a population of just over 100,000. And believe it or not, Charleston was founded on the ideals of religious tolerance. A whole host of Europeans, each experiencing a different brand of religious persecution in the seventeenth and eighteenth centuries, took up residence in the historic city and decided to get along.

Charleston is the unofficial capital of America's *Lowcountry*—the Atlantic Coastal Plain that was at various times in geological history part of the ocean floor (the soil of my garden is 100 percent sand). Already I've experienced the effects of climate change here when Hurricane Joaquin delivered a *rain bomb* (a term new to meteorology) and twenty-four inches of water fell from the sky within a day or two. My house sits sixty-three feet above sea level so it did OK. But most of the Lowcountry is, well, *low*. Rain quickly overwhelmed older infrastructure, transformed neighborhoods into lakes, washed away roadbeds, shut down traffic and cost lives. People boated down the centerline of Charleston's streets.

As beautiful and charming as Charleston is, it is also a city in grave danger. The coastal town is already committed or "locked into" irreversible flooding due to sea level rise (according to a National Academy of Sciences study released in November, 2015[52]). Twenty-five percent of Charleston's land mass will be underwater when the full effects of the CO_2 emissions *already released into the atmosphere* swell ocean waters by more than five feet. Even if civilization were to completely stop all greenhouse gas emissions *today*, a quarter of Charleston would get swamped (there is a lag time between the release of greenhouse gases and their effects). Given the fact that society *isn't* abruptly halting emissions, researchers speculate that the best case scenario for Charleston is perhaps only half of the city will be lost. But, should the nations around the world who signed onto the COP 21 Paris agreement *exceed* the 1.5°C temperature limit, should the global community end up closer to +2°C warming, then Charleston's future is likely lost. The most recent interactive online tools wipe Charleston off the map at 2°C global temperature rise.[53]

The South Carolina Department of Natural Resources predicts the coming shift in the climate will be grave and intense for my new state. In its 2013 report, the agency describes a future wherein beachfront properties and coastal wetland ecosystems erode, aquatic species die, and algae blooms poison shellfish, making the human and wildlife food chain toxic.[54] Dead zones along the coast, they say, will wipe out coral reefs and fisheries. The department laments South Carolina's inability to plan for and implement effective adaptation measures; the report blames a void in governmental leadership and a lack of political will.

Still, I remind myself that there may be possibilities to reduce the impacts of sea level rise, protect life, and maintain the area's viability. Charleston, at least, doesn't have Miami's porous limestone base.

I chose to come to Charleston to engage on the frontline of climate opportunity. So far, I've discovered a community in desperate need of honest, open, and inclusive dialogue. The good news is there are compassionate leaders here who are working hard to transform timeworn knots of political and social dysfunction. The *Charleston Strong* campaign that percolated out of Hurricane Hugo's aftermath and gelled in the horror of the AME church shooting speaks to the locals' resiliency and grace. I am actively hopeful, regardless of the outcome, that we can bring transformation and healing to this place and to Earth.

When I came to town, I needed a spiritual community, so I started the Charleston Community of Mindful Living—a meditation group in the Thich Nhat Hanh tradition. I also started blogging about the human mind-heart-environment connection. People heard about what I was up to and invited me to help lead Buddhist retreats on the topic of Ecological Mindfulness in California, Chicago, and North Carolina. But for a time I struggled over how I might engage in advocacy again, how to directly incorporate the lessons I gleaned from the Great Atlantic Teacher in the often contentious work of creating ecological and social justice. Every time I thought I *ought* to do something, I slammed head-long into a thing I was no longer willing to do. I kept coming up short and getting confused. So, to sort out my thinking, I made a list. After my experience with the Great Atlantic Teacher, I was no longer willing to:

Make enemies

Be angry

Be ambitious

Blame others

Rush

Exaggerate or spin my message

Work long hours

Argue my case

I had become more attuned to my own bodily reactions in the last few years. Whenever I was going to argue a point, my body, in a split second, would inhale sharply and lean forward. That body-clue became a signal for me to stop, take a breath, and listen for at least twice as long as I thought I might speak. When I stopped and listened, a funny thing happened: I often ended up not saying anything at all.

On the flip side, I made a list of everything I *was* willing to do. That list included:

Make friends

Cultivate happiness and share my joy with others

Work in small and simple ways

Look deeply into the roots of conflicts (cultivate insight)

Proceed peacefully (being mindful of what's
    happening in here and out there)

Speak plainly and accurately

Play with loved ones and in community

Listen and share

I thought about returning to environmental work at a not-for-profit on the climate challenge. But my new mindful advocacy legs felt wobbly, untested. If I surrounded myself with ambitious people working to change the world without first changing themselves, I feared I'd grow forgetful and slip into old habits of pushing and doing. What's more, I couldn't think of a nonprofit organization or government agency that would be willing to hire someone who refused to get excited, separate foes from allies, or meet (arbitrary) deadlines.

Colleagues, family members, and friends encouraged me to finish the book I'd been tinkering with as I sailed, and I thought writing could be a skillful form of advocacy in its own right. And then there was the work I did with the Earth Holder Project of SnowFlower Sangha. I felt comfortable turning to people who were already working on the battlefield of human hearts and minds; the mining petition I worked on felt right. So I started to think about a kind of engagement with the world I called *Mindful Advocacy*—it was a way of being, reflecting, and acting that was one hundred percent supported by the teachings of my spiritual practice. Suddenly, rather than feeling depleted by my life as a change agent, I felt energized. Rather than being worried about outcomes, I was free to do my best and let it go.

My mind couldn't help but leap to Thay's global practice community. Plum Village is an international community, founded by Thay in the mid-1970s, that today includes more than 800 monastics and 1,300 local practice centers worldwide. Every year, Thay and the Plum Village community train tens of thousands of people in applied mindfulness. It occurred to me that the Plum Village community was *uniquely prepared* to address the climate challenge because it was a global network of mindfulness practitioners aspiring to ease suffering in the world.

Early in 2015, a group of Thay's students and I began to discuss ways we could continue Thay's teaching on Earth holding and protecting, and bring attention to the spiritual path of inner and outer sustainability. The community already had a handful of beautiful but independent initiatives popping up—blogs, the Earth Holding Weekend at Deer Park Monastery, brilliant talks that went viral, a Yahoo! group discussing climate change, and the publication of a quarterly newsletter I helped edit called *Touching the Earth*. We asked ourselves what would happen if we combined our energies. Personally, I was looking for friends curious to explore ecomindfulness who would support me in furthering my commitment to a plant-based diet and who wanted to join together in Earth-healing acts of mindful advocacy. In just a few short months, we drafted a statement of purpose and vision, found ways of communicating and making decisions as a dispersed group across the continent, assembled committees, and publically launched the Earth Holder Sangha as a Plum Village affinity group. I was honored to become the first facilitator of the group's central Care-Taking Council.

Things really got going when a delegation of Plum Village monastics was invited to attend the COP 21 talks in Paris on behalf of Thay. (You can watch a video about their experience and learn more about Earth Holder Sangha by visiting EarthHolder.org). In December 2015, the monastics of Plum Village announced that the North American Earth Holder Sangha would expand internationally, and they encouraged mindfulness practitioners around the world to organize as we had done. This is how—in less than a year—an international Earth Holder movement was born.

~~~

I don't think you have to be Buddhist to be an Earth Holder. People of all stripes can wake up to their relationship with Earth and behave in ways that love, protect, and transform suffering in whatever corner of this precious planet they live. Anyone can become aware of ecological suffering, see its causes, figure out how to stop harm, and act in ways that bring happiness and healing. Every person knows that actions that lead to suffering—now or in the future—are wrong and actions that bring peace and well-being are right. So, every last one of us is welcome and needed on this path, and I trust each of us is capable of acting skillfully from a place of insight and love.

I see Earth Holders at work everywhere and they are already making a difference. I see the couple in Wisconsin who grow their own food and put in a geothermal system to heat their home. I think about the people I know who bike to work. I find inspiration in those who take time off work to participate in rallies and demonstrations. I appreciate the man and wife who built a "green" home—one that incorporated recycled materials and a solar energy array (they sell their excess wattage back to the power company). And of course, I admire A.H. Errol Harris and his wife, Marcella.

It was January 2012 when I met with the Harrises under the trees of the Dominica Botanical Gardens' Café (*Wild Hair* sat off the coast). I wanted to talk to them because the Harrises were instrumental in launching the Dominica Sea Turtle Conservation Organization (DomSeTCo)—an island initiative that simultaneously researched and protected the local turtle populations and enhanced the quality of life and economic well-being of Dominica's beach-front communities.

Three sea turtle species nested on Dominica: Leatherbacks, Greens, and Hawksbills. The government addressed the island's joblessness and hunger with policies encouraging citizens to hunt and eat the turtles. The United Nations put a halt to the policies when it listed Green Turtles as a "threatened" species and Leatherbacks and Hawksbills as "endangered." But poaching continued.

The Harrises came up with a simple idea: teach poachers and other coastal residents how to organize, market, and lead turtle hatch expeditions for tourists; aid research by tagging turtles; and patrol beaches to keep the creatures safe during nesting season. It was a great idea except for one thing: the Harrises (and their partners at the Wider Caribbean Sea Turtle Conservation Network) worried there might be a mass slaughter when poachers realized turtles were not dangerous, as they assumed, but were in fact gentle giants. In the end, DomSeTCo leaders decided to *trust* poachers because the program—to be successful—required across-the-board understanding and integrity.

Trust worked. Today, people have new livelihoods on Dominica and poaching has dropped from twenty-five annual slaughters to only one loss in two years. The initiative, which realized economic, social, and ecological goals, is a model of sustainability. The program protects the biodiversity of the marine ecology and helps people thrive as they live within planetary boundaries.[55] To my mind, Marcella and Errol are Earth Holders. What's really remarkable to me is the Harris's insight that the present didn't have to be like the past, that those who created the most harm could transform and become fellow Earth Holders.

Today I am a pilgrim in a Holy City on the frontline of the climate crisis, with a strong aspiration to join up with fellow Earth Holders so that together we can cultivate inner and outer sustainability. I return to this life on land not as an expert with answers, but as a beginner curious about the possibilities climate change holds. Truth be told, the frontline isn't just in Charleston, it's everywhere "out there" and "in here" in the battlefield of *my* mind. Climate change is me and I am it. And because the future won't look like the past, I do my best to become a little more mindful each day about what is happening in me and in the world. I try to make friends as I go.

I can hardly wait to see what happens.

NOTES

1   Benjamin H. Straus et al, "Carbon choices determine US cities committed
     to futures below sea level," Proceedings of the National Academy of Sciences,
     November 3, 2015, accessed Feb 12, 2016, http://www.pnas.org/con-
     tent/112/44/13508.full.pdf

2   "Third National Climate Assessment," *The US Global Change Research Program*,
     accessed January 4, 2016, http://www.globalchange.gov/ncadac

3   "Florida: Public Opinion on Climate Change," *Yale University School of Forestry
     and Environmental Studies*, accessed February 5, 2015, http://environment.yale.
     edu/climate-communication/files/Florida_Global_Warming_Opinion.pdf

4   "US Greenhouse Gas Emissions Flow Chart," *World Resources Institute*,
     accessed January 11, 2016, http://www.wri.org/resources/charts-graphs/
     us-greenhouse-gas-emissions-flow-chart

5   "Miami's Boom (And Climate Doom?)," *On Point with Tom Ashbrook*,
     accessed November 20, 2015, www.onpoint.wbur.org/2015/11/19/
     Miami-beach-real-estate-climate-change

6   "Miami's Boom (And Climate Doom?)," *On Point with Tom Ashbrook*,
     accessed November 20, 2015, www.onpoint.wbur.org/2015/11/19/
     Miami-beach-real-estate-climate-change

7   "Miami's Boom (And Climate Doom?)," *On Point with Tom Ashbrook*,
     accessed November 20, 2015, www.onpoint.wbur.org/2015/11/19/
     Miami-beach-real-estate-climate-change

8   Stephen C. Risner and M. Susan Lozie, "New Simulations Question the Gulf
     Stream's Role in Tempering Europe's Winters," *Scientific American*, February 1,
     2013, accessed January 6, 2016, http://www.scientificamerican.com/article/
     new-simulations-question-gulf-stream-role-tempering-europes-winters/

9   Ronald Inglehart et al., "Development, Freedom, and Rising Happiness: A Global
     Perspective (1981-2007)," (*Association for Psychological Science*, Vol 3, No. 4, 2008)
     accessed September 29, 2011, http://www.worldvaluesurvey.org/wvs/articles/
     folder_published/article_base122/files/RisingHappinessPPS.pdf

10   Richard Easterlin, "Explaining Happiness," (LA, CA, *University of Southern
     California*, 2003), this contribution is part of the special series of the Inaugural
     Articles by the members of the National Academy of Sciences elected
     April 30, 2002, accessed September 29, 2011, http://www.pnas.org/con-
     tent/100/19/11176.full.pdf

11   Eric Weiner, *The Geography of Bliss: One Grump's Search for the Happiest Places in the World* (New York: Twelve Hachette Book Group USA, 2008), Kindle edition.

12   Morbidity and Mortality Weekly Report (MMWR), "QuickStats: Number of Deaths from 10 Leading Causes—National Vital Statistics System, United States, 2010," *Center for Disease Control and Prevention*, published March 1, 2010, accessed June 6, 2016, https://www.cdc.gov/mmwr/preview/mmwrhtml/mm6208a8.htm

13   "Overweight and Obesity Statistics," National Institute of Diabetes and Digestive and Kidney Disease, accessed June 6, 2015, http://www.niddk.nih.gov/health-information/health-statistics/Pages/overweight-obesity-statistics.aspx

14   Ana Swanson, "The US isn't the fattest country in the world—but it's close." *The Washington Post*, published April 22, 2015, accessed June 6, 2016, https://www.washingtonpost.com/news/wonk/wp/2015/04/22/youll-never-guess-the-worlds-fattest-country-and-no-its-not-the-u-s/

15   Tyjen Tsai and Paola Scommegna, "US Has World's Highest Incarceration Rate," Population Reference Bureau, accessed June 6, 1026, http://www.prb.org/Publications/Articles/2012/us-incarceration.aspx

16   "The Prison Crisis," *American Civil Liberties Union*, accessed June 6, 2016, https://www.aclu.org/prison-crisis

17   "Marriage and divorce," American Psychological Association, Accessed June 6, 2016, http://www.apa.org/topics/divorce/

18   "National Statistics," National Coalition Against Domestic Violence, accessed June 6, 2016, http://www.ncadv.org/learn/statistics

19   Michele C. Black, et. al., "The National Intimate Partner and Sexual Violence Survey: 2010 Summary Report, National Center for Injury Prevention and control Centers for Disease Control and Prevention, November 2011, accessed June 6, 2016, http://www.cdc.gov/violenceprevention/pdf/nisvs_report2010-a.pdf

20   Erin Grinshteyn, David Hemenway, "Violent Death Rates: The US Compared with Other High-income OECD Countries, 2010," American Journal of Medicine, March 16 Volume 129, Issue 3, Pages 266-273, accessed June 6, 2016, http://www.amjmed.com/article/S0002-9343(15)01030-X/fulltext

21   "Facts About Alcohol," National Council on Alcoholism and Drug Dependence, Inc, accessed June 6, 2016, https://www.ncadd.org/about-addiction/alcohol/facts-about-alcohol

22   "Facts and Statistics: Drug Adiction," Michaels House, accessed June 6, 2016, http://www.michaelshouse.com/drug-addiction/the-statistics/

23 "Drug Facts: Nationwide Trends," National Institute on Drug Abuse, revised June 2015, accessed June 6, 2016, https://www.drugabuse.gov/publications/drugfacts/nationwide-trends

24 Susan Emmerich and Jeffrey Pohorski, "When Heaven Meets Earth: How a Faithful Few Changed Everything," *A Skunkfilms Production*, accessed January 6, 2016, http://whenheavenmeetsearth.org/

25 Thich Nhat Hanh, "Five Mindfulness Trainings," Plum Village Practice Center, accessed January 7, 2016, http://plumvillage.org/mindfulness-practice/the-5-mindfulness-trainings/

26 Coral Davenport and Campbell Robertson, "Resettling the First American 'Climate Refugees,'" (The New York Times, May 2, 2016) accessed May 3, 2016, http://www.nytimes.com/2016/05/03/us/resettling-the-first-american-climate-refugees.html?emc=eta1&_r=0

27 "1987 Whittier Narrows Earthquake," *Wikipedia*, accessed May 15, 2016, https://en.wikipedia.org/wiki/1987_Whittier_Narrows_earthquake

28 Brian Swimme and Thomas Berry, *The Universe Story: From the Primordial Flaring Forth to the Ecozoic Era, a Celebration of the Unfolding of the Cosmos* (New York, NY: Harper Collins Publishers, 1992), 3

29 Joanna Macy and Chris Johnstone, *Active Hope How to Face the Mess We're in without Going Crazy* (Novato, CA: New World Library, 2012) Kindle Edition.

30 Thich Nhat Hanh, *Being Peace* (Berkeley, California: Parallax Press, 1987), 68-69.

31 Ven. Bhikku Bodhi, "The Four Noble Truths of the Climate Crisis," US Buddhist Leaders Conference at George Washington and the White House, Washington, DC on May 14 2015, accessed September 19, 2015, https://www.youtube.com/watch?v=b7FViHm1xdU.

32 Shannon Hall, "Exxon Knew about Climate Change Almost 40 Years Ago," Scientific American, October 26, 2015, accessed January 13, 2016, http://www.scientificamerican.com/article/exxon-knew-about-climate-change-almost-40-years-ago/

33 Ernest Zebrowski Jr, *The Last Days of St. Pierre: The Volcanic Disaster That Claimed 30,000 Lives* (Brunswick: Rutgers University Press, 2002), Kindle Edition.

34 Carl Sagan, *Cosmos* (New York, Random House Publishing, 1980)

35 "Sand Mining," Barbudaful, accessed November 8, 2015, http://www.barbudaful.net/natural-barbuda/issues/sand-mining.html

36   Desmond Brown, "Facing Tough times, Barbuda Continues Sand Mining despite Warnings," Inter Press Service News Agency, published June22, 2013, accessed November 8, 2015, http://www.ipsnews.net/2013/06/facing-tough-times-barbuda-continues-sand-mining-despite-warnings/

37   Aldo Leopold, *A Sand County Almanac* (New York, Ballantine Books, 1966)

38   Desmond Brown, "Tiny Barbuda Grapples with Rising Seas," Inter Press Service News Agency, accessed November 12, 2015, http://www.ipsnews.net/2014/06/tiny-barbuda-grapples-with-rising-seas/

39   Desmond Brown, "Facing Tough Times, Barbuda Continues Sand Mining Despite Warnings," Inter Press Service News Agency, accessed November 16, 2016, http://www.ipsnews.et/2013/06/facing-tough-barbuda-continues-sand-mining-despite-warnings/

40   Thich Nhat Hanh, *The World We Have: A Buddhist Approach to Peace and Ecology*, (Berkeley, CA: Parallax Press, 2008), 8.

41   SnowFlower Sangha et al, "Bring Ethics to Wisconsin Mining," Change.Org, accessed May 26, 2016, https://www.change.org/p/wi-governor-candidates-bring-ethics-to-wisconsin-mining

42   Chris Mooney and Joby Warrick, "How tiny islands drove huge ambition at the Paris climate talks," The Washington Post, December 12, 2015, accessed May 26, 2016, https://www.washingtonpost.com/news/energy-environment/wp/2015/12/11/how-tiny-islands-drove-huge-ambition-at-the-paris-climate-talks/

43   NRC (2011). *America's Climate Choices: Final Report*, National Research Council. The National Academies Press, Washington, DC, USA.

44   Intergovernmental Panel on Climate Change, Fourth Assessment Report (2007), accessed May 29, 2016, https://www.ipcc.ch/publications_and_data/publications_ipcc_fourth_assessment_report_synthesis_report.htm

45   "US Greenhouse Gas Emissions Flow Chart," *World Resources Institute*, accessed January 11, 2016, http://www.wri.org/resources/charts-graphs/us-greenhouse-gas-emissions-flow-chart

46   "Global Livestock Counts, Counting Chickens," The Economist, posted July 27, 2911, accessed May 29, 2016, http://www.economist.com/blogs/dailychart/2011/07/global-livestock-counts

47   Eric Holthaus, "Our Planet's Temperature Just Reached a Terrifying Milestone," posted March 12, 2016, accessed May 29, 2016, http://www.slate.com/blogs/future_tense/2016/03/01/february_2016_s_shocking_global_warming_temperature_record.html

48   Pope Francis, "Encyclical Letter Laudato Si' of the Holy Father Francis On Care for our Common Home" *Libreria Editrice Vaticana*, accessed January 11, 2016, http://w2.vatican.va/content/francesco/en/encyclicals/documents/papa-francesco_20150524_enciclica-laudato-si.html

49   Justin Gillis, "Short Answers to Hard Questions About Climte Change," *New York Times*, posted November 28, 2015, accessed February 22, 2016, http://www.nytimes.com/interactive/2015/11/28/science/what-is-climate-change.html?_r=0

50   James Gerken, "Climate Change is a Growing National Security Concern, Say Retired Military Leaders, *Huffpost Green*, posted May 14, 2014, accessed May 29.2016, http://www.huffingtonpost.com/2014/05/14/climate-change-national-security_n_5323148.html

51   Will Steffen, et al., "Planetary boundaries: Guiding human development on a changing planet," *Science*, February 13, 2015, accessed January 11, 2014, http://www.sciencemag.org/content/347/6223/1259855.abstract

52   Benjamin H. Straus et al, "Carbon Choices determine US cities committed to futures below sea level," Proceedings of the National Academy of Sciences, November 3, 2015, accessed February 12, 2016, http://www.pnas.org/content/112/44/13508.full.pdf

53   EcoWatch, "Interactive Map Shows 414 US Cities Already Locked into Catastrophic Sea Level Rise," accessed January 16, 2015, http://ecowatch.com/2015/10/14/map-sea-level-rise/

54   South Carolina Department of Natural Resources, "Climate Change Impacts to Natural Resources in South Carolina," accessed July 6, 2016, http://www.dnr.sc.gov/pubs/CCINatResReport.pdf

55   "Action Plan for a Sea Turtle Conservation and Tourism Initiative in Dominica, 2008" accessed December 26, 2015, http://www.widecast.org/Resources/Docs/Dominica_Sea_Turtle_Action_Plan_2008.pdf

SUGGESTED RESOURCES

To cultivate ecological mindfulness and inspire loving and effective climate action, members of the Earth Holder Sangha in the Plum Village Tradition offer this list of books, films, organizations, and news sources.

Non-Fiction Books

Diane Ackerman, *The Human Age: The World Shaped by Us* (W.W. Norton and Company, 2014)

Edited by Allan Hunt Badiner, *Mindfulness in the Marketplace: Compassionate Responses to Consumerism* (Parallax Press, 2002)

Gerardo Ceballos and Anne H. Ehrlich *Annihilation of Nature: A clarion call for engagement and action to stop the 6th great extinction* (Johns Hopkins University Press, 2015)

Jonathan M. Cullen and Julian M. Allwood, *Sustainable Materials with Both Eyes Open: Without the Hot Air* (UIT Cambridge, 2012)

Jared Duval, *The Next Generation Democracy* (Bloomsbury, 2010)

Charles Einstein, *The More Beautiful World Our Hearts Know is Possible: Sacred Activism* (North Atlantic Books, 2013)

Duane Elgin, *Voluntary Simplicity* (Harper, 2010)

Daniel C. Esty and Andrew S. Winston, *Green to Gold: How Smart Companies Use Environmental Strategy to Innovate, Create Value, and Build Competitive Advantage* (Yale University Press, 2006)

Herbert Girardet and Miguel Mendonca, *A Renewable World: Energy, Ecology, Equality* (Green Books, 2009)

Paul Gilding, *The Great Disruption: How the Climate Crisis will Change Everything for the Better* (Bloomsbury Press, 2011)

Thich Nhat Hanh, *Love Letter to the Earth* (Parallax Press, 2013)

Thich Nhat Hanh, *The World We Have* (Parallax Press, 2008)

Dr. James Hansen, *Storms of My Grandchildren: Descriptions of the coming climate catastrophe and our last chance to save humanity* (Bloomsbury USA, 2010)

Paul Hawken and Amory Lovins, *Natural Capitalism: Creating the Next Industrial Revolution* (Little, Brown and Co, 2010)

Mark Hertsgaard, *Hot: Living Through the Next Fifty Years on Earth* (Mariner Books, 2012)

Randolph Hester, *Design for Ecological Democracy* (MIT Press, 2006)

Naomi Klein, *This Changes Everything: Capitalism vs. the Climate* (Simon and Schuster, 2015)

Elizabeth Kolbert *The Sixth Extinction: An Unnatural History* (Henry Holt & Co, 2014)

David Korten, *The Great Turning: From Empire to Earth Communities* (Kumarian Press, 2007)

Anna Lappe, *Diet for a Hot Planet: The Climate Crisis at the End of Your Fork and What You Can do About It* (Bloomsbury, 2010)

Frances Moore Lappe, *EcoMind: Changing the Way We Think to Create the World We Want* (Nation Books, 2011)

John Daido Loori, *Buddhism and Ecology: The Interconnection of Dharma and Deeds* (Harvard University Press, 1998)

John Daido Loori, *Teachings of the Earth: Zen and the Environment* (Shambhala Publications, 1999)

Richard Louv, *The Last Child in the Woods: Saving our Children from Nature-Deficit Disorder* (Algonquin Books, 2008)

Amory Lovins and Rocky Mountain Institute, *Reinventing Fire: Bold Business Solutions for the New Energy Era* (Chelsea Green Publishing, 2013)

Joanna Macy, *World as Lover, World as Self* (Parallax Press, 1991)

Joanna Macy and Molly Young Brown, *Coming Back to Life* (New Society Publishers, 1998)

Joanna Macy and Chris Johnstone, *Active Hope: How to Face the Mess We're in without Going Crazy* (New World Library, 2012)

Joel Makower, *Strategies for the Green Economy: Opportunities and Challenges in the New World of Business* (McGraw-Hill, 2008)

William McDonough, *Upcycle: Beyond Sustainability: Designing for Abundance* (North Point Press, 2013)

Bill McKibben, *Eaarth: Making a Life on a Tough New Planet* (Times Books, 2010)

Bill McKibben, *Oil and Honey: The education of an unlikely activist* (St Martin's Griffin, 2014)

Zachia Murray, *Mindfulness in the Garden: Zen Tools for Digging in the Dirt* (Parallax Press, 2012)

Christian Parenti, *Tropic of Chaos: Climate Change and the New Geography of Violence* (Nation Books, 2012)

Mary Pipher, *The Green Boat: Reviving Ourselves in Our Capsized Culture* (Riverhead Trade, 2013)

Gary Snyder, *The Practice of the Wild* (Counterpoint; expanded edition August 17, 2010)

Edited by John Stanley, David R. Loy, and Gyurme Dorje, *A Buddhist Response to THE CLIMATE EMERGENCY* (Wisdom Publications, 2009)

Sustainable World Coalition, *Sustainable World Sourcebook: Critical issues, Inspiring Solutions, Resources for Action, the Essential Guidebook for the Concerned Citizen* (Vinti Allen, 2014)

Edited by Llewelly Vaughan-Lee, *Spiritual Ecology: The Cry of the Earth* (the Golden Sufi Center, 2013)

Jay Walljasper, *All that We Share* (New Press, 2010)

Edited by Claude Whitmyer, *Mindfulness and Meaningful Work: Explorations in Right Livelihood* (Parallax Press, 1994)

Meg Wheatley, *So Far From Home: Lost and Found in Our Brave New World* (Berrett-Koehler Publishers, 2012)

Andrew Winston, *The Big Pivot: Radically Practical Strategies for a Hotter, Scarcer, and More Open World* (Harvard Business Review Press, 2014)

Documentaries/Films/Video Presentations

Animals and the Buddha, a 2015 documentary by Dharma Voices for Animals, https://www.youtube.com/watch?v=NlLUPzMi0Rw

Bidder 70, 2013, produced and directed by Beth Gage and George Gage, http://www.bidder70film.com/#!about/cee5

The Case for Optimism on Climate Change with Al Gore, a February 2016 TED Talk, http://www.ted.com/talks/al_gore_the_case_for_optimism_on_climate_change?utm_source=newsletter_weekly_2016-02-27&utm_campaign=newsletter_weekly&utm_medium=email&utm_content=talk_of_the_week_image

Call to Earth : A Message from the World's Astronauts to COP21. 2015, by Planetary Collective, https://www.youtube.com/watch?v=NN1eSMXI_6Y

COP21: His Holiness the Dalai Lama's Message. 2015 by Tibet TV, https://www.youtube.com/watch?v=iYBMLsc64HM

Cowspiracy: The Sustainability Secret, a 2014 documentary by directors Kip Andersen and Keegan Kuhn, http://www.imdb.com/title/tt3302820/

The Crisis of Civilization, a 2015 documentary directed by Dean Pocket, http://systemchangenotclimatechange.org/article/crisis-civilization-full-length-documentary-movie-hd

David Steindl-Rast: Want to be happy? Be grateful, a 2013 Ted Talk, https://www.youtube.com/watch?v=UtBsl3j0YRQ

Disruption, a 2014 documentary directed by Kelly Nyks and Jered P Scott, https://vimeo.com/105412070

Earth 2050 the Future of Energy, a 2015 documentary by director Lilibeth Foster, https://www.youtube.com/watch?v=TaSpGbaYQRs

The Economics of Happiness, a 2013 documentary by Helene Norberg-Hodge, Steven Gorelick, and John Page, https://vimeo.com/ondemand/theeconomicsofhappiness

Gasland, a 2010 documentary by director Josh Fox, http://www.imdb.com/title/tt1558250/

Global Warming: The Signs and the Science, a 2012 PBS report hosted and narrated by Alanis Morissette, https://www.youtube.com/watch?v=xVQnPytgwQ0

Global Warming: What You Need To Know, with Tom Brokaw, a 2012 report by Discovery Channel, https://www.youtube.com/watch?v=xcVwLrAavyA

Home, a 2009 documentary by director Yann Arthus-Bertrand, http://www.imdb.com/title/tt1014762/

Hope in a Changing Climate, a 2009 BBC documentary with John D. Liu, https://www.youtube.com/watch?v=bLdNhZ6kAzo

How to Save the World with Johan Rockstrom, a 2015 presentation by the executive director of the Stockholm Resilience Center, http://www.huffingtonpost.com/entry/if-you-want-to-save-the-planet-watch-this-video_us_560d4dcfe4b0dd85030af4b9

An Inconvenient Truth, a 2006 documentary by director Davis Guggenheim, http://putlocker.is/watch-an-inconvenient-truth-online-free-putlocker.html

Inner Peace to World Peace, a 2012 film by Uplift media channel, http://upliftconnect.com/peace-day-video/

Planetary, 2016, directed by Guy Reid, http://weareplanetary.com/

Plum Village Presence at The Paris Climate Conference, 2015 Plum Village Online, https://www.youtube.com/watch?v=D8qcZ14jTMw

Pope Francis and the Environment: Why His New Climate Encyclical Matters, 2015 panel discussion at Yale Univeristy, https://www.youtube.com/watch?v=qtsAAvqavBg

Pope Francis' speech to UN loaded with social issues, 2015 Rome Reports, https://www.youtube.com/watch?v=EghYFUM94Ho

Racing Extinction, a 2015 documentary by director Louie Psihoyos, http://www.amazon.com/Racing-Extinction-Louie-Psihoyos/dp/B0184RE1TG/

Mr. Ray Anderson: The Business Logic of Sustainability, a 2009 Ted Talk, https://www.ted.com/talks/ray_anderson_on_the_business_logic_of_sustainability?language=en

Ron Finley: A guerilla gardener in South Central LA, a 2013 TED Talk, https://www.youtube.com/watch?v=EzZzZ_qpZ4w

Sacred Economics with Charles Einstein a 2102 film by director Ian MacKenzie, https://www.youtube.com/watch?v=EEZkQv25uEs

The Scientific Case for Urgent Action to Limit Climate Change, a 2013 scientific presentation by Richard CJ Somerville at the Scripps Institute of Oceanography, http://www.ucsd.tv/search-details.aspx?showID=24910

The Story of Earth and Life, a2011 documentary by National Geographic, https://www.youtube.com/watch?v=57merteLsBc

Thich Nhat Hanh Address to the Parliament of World Religions, December 2009, https://vimeo.com/8184071

This Changes Everything, a 2015 documentary by director Avi Lewis and writer Nomi Klein, http://www.imdb.com/title/tt1870548/

Uncommon Conversations: Thich Nhat Hanh and Br. David Steindl-Rast and Gratefulness, a 2015 interview directed by Nicholas Weidner, https://m.youtube.com/watch?v=AZKsOfYURtI

Organizations

350.org, "Climate-focused campaigns, projects and actions led from the bottom-up by people in 188 countries," www.350.org

Bioneers "Revolution from the Heart of Nature," www.bioneers.org

Buddhist Peace Fellowship "Cultivating Compassionate Action," www.budhistpeacefellowship.org

Buddhist Climate Action Network "Taking action on climate with compassion for all living beings," www.globalcan.org

The Center for Resilient Cities "Supporting healthy resilient people in healthy resilient places," www.resilientcities.org

Citizen's Climate Lobby "A nonprofit, nonpartisan grassroots advocacy organization focused on national policies to address climate change," citizensclimatelobby.org

Climate One "Changing the conversation about energy, the economy, and the environment," www.ClimateOne.org

The Climate Reality Project "Catalyzing a global solution to the climate crisis by making urgent action a necessity across every level of society," www.climaterealityproject.org

Divest Invest "Calling on investors of every stripe to divest from the fossil fuel industries deepening the climate crisis and invest in climate solutions," divestinvest.org

EcoAmerica Moment US "Grows the base of popular support for climate solutions in America," ecoamerica.org/momentus

Energy and Enterprise Initiative "A nonprofit educational effort at George Mason University committed to building public understanding of free enterprise and its promise to solve energy and climate challenges," republicen.org

Evangelical Environmental Network "Educates, inspires, and mobilizes Christians in their effort to care for God's creation... and advocates for actions and policies that honor God and protects the environment," creationcare.org

The International Earth Holder Sangha "An affinity group in the Plum Village Tradition awakening a great togetherness to bring transformation and healing to earth," www.EarthHolder.org

Interfaith Power & Light "A religious response to global warming," www.interfaithpowerandlight.org

One Earth Sangha "Expressing a Buddhist response to climate change and other threats to our home," www.oneearthsangha.org

The Pachamama Alliance "Weaving indigenous wisdom and modern knowledge for a thriving, just, and sustainable world," www.pachamama.org/about

Plum Village Mindfulness Practice Center " The first monastic community founded by Zen Master Thich Hnat Hanh in the West," plumvillage.org

Progressive Christians Uniting "Engaging people and communities to embody Jesus' way of compassion and justice for our world," www.progressivechristiansuniting.org

The Racing Extinction Challenge a tool kit based on Racing Extinction: The Film, racingextinction.com

Rainforest Action Network "Campaigns for the forests, their inhabitants and the natural systems that sustain life," www.ran.org

Respectful Revolution "A national, not-for-profit advocacy project seeking to document positive action and inspire change," www.respectfulrevolution.org

Right Livelihood Award "The alternative Nobel Prize to those offering practical and exemplary answers to the most urgent challenges facing us today" www. rightlivelihood.org

Sustainable World Coalition "Individuals and organizations working toward a world that is environmentally sustainable, socially just and spiritually fulfilling," www. SWCoalition.org

Transition Town Movement "A grassroots initiative seeking to build community resilience in the face of peak oil, climate change and the economic crisis," transitionus.org

Seed Freedom "Protecting the biodiversity of the planet by defending the freedom of the seed to evolve with integrity, self-organization, and diversity," www. vandanashiva.org

Wider Caribbean Sea Turtle Conservation Network "Realizing a future where all inhabitants of the Wider Caribbean Region, human and sea turtle alike, can live together in balance," www.widecast.org

Women's Earth & Climate Action Network International, "Engaging women worldwide to take action as powerful stakeholders in climate change and sustainability solutions," wecaninternational.org

World Resources Institute (WRI) "Advancing transformative solutions for climate change mitigation and adaptation," www.wri.org/our-work/topics/climate

World Wildlife Fund (WWF) "Create a climate-resilient and zero-carbon world, powered by renewable energy," www.worldwildlife.org/initiatives/climate, and www.worldwildlife.org/industries/dairy

Climate Information Sources

Climate Central a science and news organization, www.climatecentral.org

Climate Progress thinkprogress.org/climate/issue

Climate Voices a network that brings nonpartisan conversations about the research findings of the majority of climate scientists to citizens, climatevoices.org

Forum on Religion and Ecology at Yale exploring religious world views, texts, ethics, and practices in order to broaden understanding of the complex nature of current environmental concerns, fore.yale.edu/publications/newsletters

Fossil Free Funds an online platform to search and compare mutual funds, fossilfreefunds.org/funds

Huffington Post www.huffingtonpost.com/green and www.huffingtonpost.com/impact

Inside Climate News A Pulitzer Prize-winning, nonprofit, nonpartisan news organization dedicated to covering climate change, energy, and the environment," www.InsideClimateNews.org

Katharine Hayhoe A Climate Scientist's Blog, www.Katharinehayhoe.com

New York Times www.nytimes.com/pages/science/earth/index.html

Policy Solutions A free and open sourced energy polity simulator, www.energypolicy. solutions/

United Nations Environmental Programme (UNEP) "The leading global environmental authority that sets the global environmental agenda," www.unep. org/climatechange/

DESMOG "Clearing the PR pollution that clouds climate science," www.desmogblog. com

UN Climate Change Newsroom The United Nations Framework Convention on Climate Change, newsroom.unfccc.int and newsroom.unfccc.int/1758.aspx

US SIF the forum for sustainable and responsible investment, charts.ussif.org/mfpc

GRATITUDES

Ocean of Insight is made of many nonbook elements: people, animals, plants, and Earth herself. I thank the cosmos for all that cradles and sustains me and mention here the relative few who gave *Ocean of Insight* its final nudge into being.

First, I am grateful to my Zen Master, the Venerable Thich Nhat Hanh, and to the four-fold Plum Village Community. Monastics, I deeply appreciate your friendship and skillfulness in transmitting ancient teachings; they bring me tremendous happiness and freedom. I thank the Plum Village lay community too—especially my spiritual mentor, teacher, and fellow sailor Cheri Maples, along with my friends in SnowFlower Sangha and the Charleston Community of Mindful Living. You help me find wisdom in the midst of daily life.

To members of the Plum Village Earth Holder Sangha—especially John Bell, Andrew Deckert, Jerome Freedman, John Freese, Rev. Jayna Geiber, Nomi Green (my fellow ecochaplaincy grad!), George Hoguet, Laura Hunter, Brian Kimmel, Jack Lawlor, Keith Miller, Kenley Neufeld, and Joyce Singh—thank you for your friendship, energy, insight, and contributions to this book's Suggested Resources.

I am deeply grateful to the men and women of the Center for Resilient Cities (www.ResilientCities.org) for being my early and steadfast partners in environmental justice. Chief among those I'm indebted to are Nancy Frank, Joe Sensenbrenner, Geri Weinstein-Breunig, and Anne Whalen, because you endowed me with your strength and know-how and kept me laughing throughout. Thank you for your book critiques and encouragements, too. I hope I have returned to you at least a fraction of your love and support.

I must also say thank you to the people who shared their love of sailing. Don Cannon, I appreciate your putting the jib sheet into my hand when

I was fifteen years old. I am grateful to Peter Grimm (Doyle Sailmakers); Tom Harney (Jordon Yachts); Tom Holland (Holland Marine); Wayne Horne (Boat Doctor); Captain John Kelly (Marine Service), Chris Parker (Marine Weather Center); and Lester Forbes (Forbes Electronics) for services vital to Dave, me, and our ship *Wild Hair*. Thanks also to the women sailors who—by their life's example—taught me that sailing is (also) a woman's sport. In particular, I want to express gratitude toward double circumnavigator and author Beth Leonard and WomenandCruising.com cofounders Gwen Hamlin, Kathy Parsons, and Pam Wall.

Also, here's a shout-out to Elaine Lembo and Mark Pilsbury at *Cruising World* magazine and George Day and the crew at *Blue Water Sailing Magazine* for publishing early versions of several of my sailing yarns. It was you who made me believe others might benefit from knowing my sailing mishaps and insights.

I am indebted to our fellow cruisers, principally: Carl and Carrie Butler aboard *s/v Sanctuary*, Rich and Margaret Escoffery on *s/v Dance Aweigh*, Fort and Michelle Felker aboard *s/v Carousel*, Dennis Jay and Julia Newhouse on *s/v Delta Blue*, John and Alex aboard *s/v Free Spirit*. Thank you for being terrific company and teaching me volumes.

I appreciate very much the men and women of the US Coast Guard and the Bahama Air and Sea Rescue Association for protecting the lives and property of mariners, listening in to my Mayday call on that bleak night, and taking action to save Dave's life. You are heroes.

Along the way, it was my extreme privilege to interview professional environmental colleagues up and down the Caribbean island chain. Cyrille Barnerias from a Ministry of Environment (Martinique); Christine Finney M.Sc. from EcoDive (Grenada); Marcella and Errol Harris from DomSeTCO (Commonwealth of Dominica); Crafton J. Isaac from the Ministry of Agriculture, Forestry, and Fisheries (Grenada); J. Bishnu Tulsie from the Saint Lucia National Trust (Saint Lucia)—thank you for helping me understand your island societies, local ecologies, and heart's concerns.

Hooray also for the powerhouse talents at the University of Wisconsin Writers' Institute, particularly: Christine DeSmet, Laurie Sheer, and Amy Lou Jenkins. I appreciate your critiques, encouragements, and introduction to the professional world of creative nonfiction. Thanks also to my longtime friend and librarian Ronda Kucher for giving *Ocean of Insight* its structure.

Of course, I am *incredibly* grateful to the entire team at Parallax Press, including Rachel Neumann, Nancy Fish, Terri Saul, Earlita Chenault, designer Debbie Berne, illustrator John Barnett, and my supremely gifted and supportive editor, Jennifer Kamenetz. I feel fortunate to have benefitted from your collective genius. Thank you for believing in this book and helping it shine.

Without the love and support of my family, *Ocean of Insight* wouldn't exist. Mom and Dad, thank you for giving me curiosity and a solid foundation from which to springboard. Eland and Maggie, I appreciate more than you could know that you were independent and trustworthy young adults so your dad and I could go adventuring. Brother Robb, thank you for a lifetime of giggles and for being a supportive reader (and yes, the book really is all about you!). Khaliun, thank you for nourishing me with weeks of delicious meals during my editing crunch. Family is my snug anchor in the storm.

Last but not least, I thank my fellow *Wild Hair* crewmembers. Dinghy the Sailor Cat, you are an awfully good sport and it was great of you to (mostly) keep your sailing criticisms to yourself. So much common-sense at sea came from the question: what would Kitty do? Finally, my dear Dave, you are the center of my universe. Please reread chapter six to get an inkling of the gratefulness I feel for you. Thank you for a lifetime of love, companionship, and wacky escapades that predate and follow our sailing odyssey. I'm so very happy to be together on this wild ride.

ABOUT THE AUTHOR

Heather Lyn Mann loves this precious planet. Environmental advocacy, blue water sailing, and eco-mindfulness teaching permeate her life and writing as expressions of her adoration. Zen Master Thich Nhat Hanh ordained Heather in 2006 into his core community of practice—the Order of Interbeing. Currently, she is facilitator of the North American Earth Holder Sangha in the Plum Village Tradition. Heather offers Keynotes and seminars on a spiritual response to climate change, Earth as teacher and mindfulness. Follow her on Twitter @HeatherLynMann, Facebook/HeatherLynMann, or visit her online at HeatherLynMann.com.

ABOUT THE INTERNATIONAL EARTH HOLDER SANGHA

The Earth Holder Sangha* is an affinity group within the Plum Village International Community of Engaged Buddhists founded by Zen Master Thich Nhat Hanh. Organized in North America in 2015, the Earth Holder Sangha aims to awaken a great togetherness to bring transformation and healing to Earth. We are guided by the engaged Buddhist ethics of the Five and Fourteen Mindfulness Trainings.

For all who love and care for the Earth, who aspire to deepen their connection with the Earth, and who want to contribute to a collective societal awakening, we:

Nourish a community of "Plum Village Earth Holders" with mindfulness practices and teachings that explore the suffering and transformation of suffering relating to climate change and inequity

Engage and link Mindful Right Actions locally, nationally, and internationally to heal and transform climate change and inequity

Enhance the sustainability of our Plum Village monasteries, lay practice centers, and local sanghas as models for an alternative future

LEARN MORE

Visit the Earth Holder Sangha website and sign up for the quarterly *Touching the Earth* newsletter at EarthHolderSangha.org.

Like us on Facebook.com/EarthHolderSangha

Follow us at Twitter.com/EarthHolderPV

Read the Earth Holder Sangha founding documents at http://bit.ly/2b96StD

Do you have questions or ideas? Contact us at Inquiries@EarthHolderSangha.org.

A portion of the proceeds from your book purchase supports Thich Nhat Hanh's peace work and mindfulness teachings around the world. For more information on how you can help, visit www.thichnhathanhfoundation.org. Thank you.

---

* The Bodhisattva Dharanimdhara is described in Chapter 25 of The Lotus Sutra, an important Mahayana Buddhist text. Dharanimdhara means "to hold, protect, or preserve the Earth." Zen Master Thich Nhat Hanh translated the name of the Bodhisattva Dharanimdhara as "Bodhisattva Protector of the Earth" or "Earth Holder," and has written about Earth Holding in his books *The World We Have*, *Peaceful Action, Open Heart*, and *Love Letter to the Earth*..

PARALLAX PRESS

Parallax Press is a nonprofit publisher, founded and inspired by Zen Master Thich Nhat Hanh. We publish books on mindfulness in daily life and are committed to making these teachings accessible to everyone and preserving them for future generations. We do this work to alleviate suffering and contribute to a more just and joyful world.

For a copy of the catalog, please contact:
Parallax Press
P.O. Box 7355
Berkeley, CA 94707
Tel: (510) 540-6411
parallax.org